/S

LANDSCAPE CHARACTER:

PERSPECTIVES ON MANAGEMENT AND CHANGE

THE NATURAL HERITAGE OF SCOTLAND

Each year since it was founded in 1992, Scottish Natural Heritage has organised or jointly organised a conference that has focused attention on a particular aspect of Scotland's natural heritage. The papers read at the conferences, after a process of refereeing and editing, have been brought together as a book. The titles already published in this series are

1. *The Islands of Scotland: a Living Marine Heritage*
    Edited by J.M. Baxter and M.B. Usher (1994), 286pp.

2. *Heaths and Moorland: Cultural Landscapes*
    Edited by D.B.A. Thompson, A.J. Hester and M.B. Usher (1995), 400pp.

3. *Soils, Sustainability and the Natural Heritage*
    Edited by A.G. Taylor, J.E. Gordon and M.B. Usher (1996), 316pp.

4. *Freshwater Quality: Defining the Indefinable?*
    Edited by P.J. Boon and D.L. Howell (1997), 552pp.

5. *Biodiversity in Scotland: Status, Trends and Initiatives*
    Edited by L.V. Fleming, A.C. Newton, J.A. Vickery and M.B. Usher (1997), 309pp.

6. *Land Cover Change: Scotland from the 1940s to the 1980s*
    By E.C. Mackey, M.C. Shewry and G.J. Tudor (1998), 263pp.

7. *Scotland's Living Coastline*
    Edited by J.M. Baxter, K. Duncan, S.M. Atkins and G.Lees (1999), 209pp.

This is the eighth book in the series.

LANDSCAPE CHARACTER: PERSPECTIVES ON MANAGEMENT AND CHANGE

Edited by Michael B. Usher

EDINBURGH: THE STATIONERY OFFICE

First published in 1999

ISBN 0 11 497266 4

British Library Cataloguing in Publication Data
A catalogue record for this book is available from the British Library.

Cover photography: Michael B. Usher

**Published by The Stationery Office Limited
and available from:**

The Stationery Office Bookshops:

71 Lothian Road, Edinburgh EH3 9AZ
0870 606 55 66 Fax 0870 606 55 88
123 Kingsway, London WC2B 6PQ
0171 242 6393 Fax 0171 242 6394
68-69 Bull Street, Birmingham B4 6AD
0121 236 9696 Fax 0121 236 9699
33 Wine Street, Bristol BS1 2BQ
0117 926 4306 Fax 0117 929 4515
9-21 Princess Street, Manchester M60 8AS
0161 834 7201 Fax 0161 833 0634
16 Arthur Street, Belfast BT1 4GD
01232 238451 Fax 01232 235401
The Stationery Office Oriel Bookshop
18-19 High Street, Cardiff CF1 2BZ
01222 395548 Fax 01222 384347

The Publications Centre
(Mail, telephone and fax orders only)
PO Box 276, London SW8 5DT
Telephone orders 0870 600 55 22
Fax orders 0870 600 55 33

Accredited Agents
(see Yellow Pages)

and through good booksellers

PREFACE

" The landscape of Scotland is rich and varied ... from the rich agricultural heartlands of Aberdeenshire to the pattern of islands and sea lochs of Argyll and Lochaber; from the dramatic high cliffs of Caithness to the flat wind blown and sand dominated lands of the Uists; from the bings of West Lothian to the gorsy knolls of Dumfriesshire; from the remote core montane massifs of Wester Ross to the wooded valleys of the Borders.

This wide range of diverse and distinct landscapes is a result of the combination of both natural and human influences on the land. The processes and patterns of landform, land cover and land use have evolved over a long time period which began over 10,000 years ago with the retreat of the last ice age.

The Scottish landscape is not static. Past changes - in particular the interaction of geology and climate - have resulted in the fundamental shape and form of the Scottish landscape. Subsequent human adaptation of the landscape to meet people's needs for food, building material, fuel and communications has given us the landscapes of today." (Anon, 1998)

Landscape was the focus of the seventh annual conference of Scottish Natural Heritage (SNH), which was held in Aberdeen jointly with the Macaulay Land Use Research Institute (MLURI). The conference developed a focus on landscape character from many points of view.

From an SNH perspective, it marked the culmination of a major research programme on landscape character assessment. This programme started in 1994 and, with 29 separate studies that cover the whole of the land and coastal area of Scotland, is virtually a 'jigsaw' that can be put together to give a national perspective. This innovative programme is described by Hughes and Buchan (this volume). The first two parts of the book, chapters 1 to 10, explore many aspects of landscape character assessment, both the inputs to it and the applications of it.

Land use is intimately linked with landscape. From a MLURI perspective, the use of land is at the interface of the physical factors that determine the shape of the land, the chemical factors that determine its fertility, the biological factors that determine the plant and animal species that live there, and the human factors that provide the cultural overlay. Many tools have been developed over the last decade, allowing both the analysis of landscapes and visualization of the effects of land use change. The third part of the book, chapters 11 to 17, explores many of these developments, though we are aware that the published word cannot do justice to the moving computer images that were demonstrated at the conference.

We wish to make two further points about the conference and the development of this book. First, time was set aside for a set of four workshops. These were lively and involved interactions between many of the people who attended the conference. Short reports on these workshops were prepared and are included as chapters 18 to 21. Second, apart from these four reports, all chapters have been commented on by two independent referees, and have been modified as a result of the referees' and editorial comments. We feel that this peer reviewing of the contents of this book is an essential feature of maintaining the quality of scientific research and review to which both MLURI and SNH aspire.

We should like to thank the two main organisers of the conference, Dr David Miller from MLURI and Ms Rebecca Hughes from SNH, who developed much of the programme. Much of the administrative support was provided by Ms Jane Lunn (MLURI) and Mrs Helen Forster (SNH), and we should like to thank them both. We are also grateful to the large number of referees who have commented on individual chapters, and to the University of Aberdeen on whose premises the conference was held and who provided the venue for the conference dinner. We are grateful to the sponsors of the event, especially Scottish Enterprise, for their contributions in both cash and kind. However, we reserve a special 'thank you' for Miss Jo Newman (SNH) who has handled all of the manuscripts, doing the detailed editorial changes, and helping to bring this book to completion in so many ways.

Finally, we should like to echo some words from the concluding chapter by Fry *et al.* (this volume). They say "We must remember that landscapes are dynamic and always changing in response to natural and anthropogenic driving forces. Mapped landscape units and their boundaries rapidly become fixed in the minds of planners and managers. The challenge will be to map landscapes in a way that allows for dynamic changes ...". The landscape of Scotland has always changed, and will continue to change, due to both natural and anthropogenic processes. In terms of the land cover, these changes over the last half century are clearly demonstrated by the National Countryside Monitoring Scheme (Mackey *et al.*, 1998). However, these changes are also affecting the landscape, and it is our opinion that both the completion of the national landscape character assessment, and the use of technology in undertaking and visualizing the interrelation of changing land use and changing landscape, are essential components of the understanding of the sustainability of Scotland's natural heritage. We are pleased that SNH and MLURI have been able to co-operate in both the conference and this book, and we look forward to developing this co-operation in the future.

T. Jeff Maxwell
Director, Macaulay Land Use Research Institute

Michael B. Usher
Chief Scientist, Scottish Natural Heritage

References

Anon. (1998). *The Landscape Character of Scotland: the National Programme of Landscape Character Assessment.* Perth, Scottish Natural Heritage.

Mackey, E.C., Shewry, M.C. and Tudor, G.J. (1998). *Land Cover Change: Scotland from the 1940s to the 1980s.* Edinburgh, The Stationery Office.

CONTENTS

Preface

List of Contributors

Foreword Lord Sewel

PART ONE: LANDSCAPE CHARACTER

1 The Landscape Character Assessment of Scotland
   Rebecca Hughes and Nigel Buchan   1

2 The Use of Landscape Character in Development Planning:
   Two Case Studies
   S.P. Bennett, L. Campbell and I. Nicol   13

3 Landscape Assessment in the Natural Heritage Zones Programme
   Frances Thin   23

4 Defining the Characteristic Landscape Attributes of Wild Land in Scotland
   Dominic Habron   34

5 Local Distinctiveness in Landscape Character
   Roland Gustavsson   41

PART TWO: APPLYING CONCEPTS OF LANDSCAPE ASSESSMENT

6 Intergrating Forest Design Guidance and Landscape Character
   Simon Bell and Nicholas Shepherd   57

7 Sustainable Tourism and the Landscape Resource: a Sense of Place
   Duncan Bryden   66

8 Assessing Public Perception of Landscape in Wales:
   a LANDMAP Approach
   J.M. Bullen   78

9 Associations Triggered by Specific Landscape Characteristics
   Caroline M. Hägerhäll   83

10 Landscape Character: its Role in Planning for New Housing in Scotland's
   Countryside
   John Moir, David Rice and Allan Watt   88

PART THREE: TECHNIQUES AND MODELS

11 Using Aerial Photography in Static and Dynamic
    Landscape Visualization
    D.R. Miller   101

12 Mapping Remote Areas Using GIS
    Steve Carver and Steffen Fritz   112

13 Approach to Landscape Character Using Satellite
    Imagery and Spatial Analysis Tools
    Hubert Gulinck, Hans Dufourmont and Ingrid Stas   127

14 High Resolution Geographic Image Data for
    Landscape Visualization
    Alun C. Jones   135

15 The Use of Information Technology in Landscape Planning
    Peter Minto   144

16 Modelling the View: Perception and Visualization
    Ian D. Bishop   150

17 Historic Land Use Assessment Project
    Piers Dixon, Lynn Dyson Bruce, Richard Hingley and Jack Stevenson   162

PART FOUR: TOPICS IN LANDSCAPE MANAGEMENT AND CHANGE: REPORTS FROM THE
          BREAKOUT SESSIONS

18 Planning and Policy Issues
    Anne Lumb and David Tyldesley   171

19 Cultural Issues
    T. Chris Smout, Charles Withers, Lesley MacInnes and Fiona Lee   176

20 Monitoring and Measuring Landscape Change
    R.V. Birnie, E.C. Mackey, S. Leadbeater and A. Bennett   180

21 Technical Developments in Typographic Data Capture, Integration,
    Manupulation and Management
    Marshall Fairbairn and Alistair Law   183

PART FIVE: A CONCLUDING PERSPECTIVE

22 Geographic Information for Research and Policy:
    a Norwegian Landscape Perspective
    Gary Fry, O. Puschmann and W. Dramstad   189

INDEX   204

LIST OF PLATES
(between pages 80 and 81)

Plate 1   Three images of Scotland's landscape, indicating a designed landscape, recreation in a lochside landscape, and a landscape portraying seasonality in the hills.

Plate 2   The level 3 map demonstrating the groups of the Scotland-wide landscape character assessment.

Plate 3   Dumfries and Galloway landscape assessment: the landscape character types.

Plate 4   The landscape character assessment of the City of Aberdeen. The landscape character areas and types are explained in the text.

Plate 5   Examples of four photographs used in the questionnaire, showing (a) coastal, (b) montane (c) native woodland, and (d) man-made artefacts.

Plate 6   Part of the Alnarp landscape laboratory; the aerial photograph, taken in 1994, shows an overview of the one kilometre long, open water-course, and its three different small water bodies. These are equal in size but with three very different designs. Beside the water-course there are some newly established meadow corridors and a newly planted multi-functional plantation.

Plate 7   Oxhagen is one of the few well-preserved grazed area in Scania, and is one of the main reference landscapes and long term research areas. Throughout history the pastoral landscape has been one of the most popular landscape types, but as such it has been viewed as very static through time. The aerial photograph, taken in 1996, gives an overview.

Plate 8   A view of Oxhagen in 1983.

Plate 9   A similar view of Oxhagen in 1998 to that in Plate 8. After an interval of fifteen years, this photograph illustrates the recent simplification processes.

Plate 10   Tjärö, an island in the south-east part of Sweden, is characterized by its landforms, rocks, grazed grasslands, and tree- and shrub-rich landscapes. It has a cultural identity with traditional buildings, fences, stone walls and pollarded trees. Tjärö is used every year for educational programmes related to landscape design and land management.

Plate 11   Sheet 10 for Upland Glens landscape character type form the Landscape Design Guidance for Forests and Woodlands in Dumfries and Galloway. The plate shows the location plan of landscape character type areas and an analysis of the existing landscape character, illustrated by a typical sketch and photographs.

Plate 12   As Plate 11, but the reverse side of Sheet 10 for Upland Glens landscape character type. The plate provides succinct design guidance describing how new woodland and forests can be integrated within the existing landscape character, illustrated both by a sketch elevation and plan view.

Plate 13   Photograph 1 – the rural village used in the first study example.

Plate 14   Photograph 2 – the wooded vale landscape also used in the first study example.

Plate 15   Original picture: characteristic view of the village of Krokshult, with red cottages, wooden fences, mounds of stones and a mosaic of small fields and pastures.

Plate 16   Manipulated picture: the red cottages and wooden fences have been erased using PhotoShop 3.0.

Plate 17   An example of a new countryside house development that is poorly located in its landscape, as explained in the text.

Plate 18   An example of a new countryside house development that is well located in its landscape, as explained in the text.

Plate 19   Illustration of the radial method. Different metrics can be derived from parameters such as radial length, nature of obstacles, sequence of lines, radial length variation, land cover types crossed by the lines, etc. The illustration is an extract of the land cover dataset of Flanders. Each pixel is 20 m square. Basic land cover classes are woodland (green), densely built (dark red), sparsely built (pink) and bare soil and low vegetation (white).

Plate 20   Visual complexity as measured by a regular sample of 64 view lines in each pixel of open space. Complexity increases with lighter shades of blue. Green patches are woodland, red patches are built up, and white lines are major roads.

Plate 21   Examples of 3D landscape visualizations that can be created today with a variety of geographic data and real landscape objects. (a) 3D rendering of 10m resolution satellite imagery over a digital terrain model using World Construction Set (© Questar Productions). (b) A 3D visualization using ERDAS VGIS with 50 cm resolution aerial photography on a 50 m DTM (© The GeoInformation Group). (c) Visualization using satellite imagery but incorporating objects, here showing the flightline of an aircraft through the model (image reproduced with permission by ERDAS UK). (d) A rendered image of the globe using low resolution (4 km) satellite data merged with a global DTM and wrapping around a sphere (© GlobalVisions images).

Plate 22   A frame from an animated drive along the Great Ocean Road, Victoria, Australia. A terrain model and the introduced objects are rendered from an identical view-point as the video frame. Terrain was rendered in black to hide the building, as appropriate, but allowing the video to project through.

Plate 23   A sample web page incorporating an interactive perception survey. Users record their assessment of visual quality and safety. When they click on location on the affect response diagram (Russell and Lanius, 1984) the survey checks that their responses are complete and valid and then moves on to the next image.

Plate 24   Aerial photograph of Parks of Aldie, Kinross, from the east, showing Aldie Castle and its policies at the bottom left, and the Improvement Period fields and plantations covering the rest of the photograph within which there are traces of relict pre-improvement settlement and rig cultivation.

Plate 25   Aerial photograph of Ettleton Churchyard and Kirkhill, Liddlesdale, Borders, from the east showing extensive pre-improvement relict landscapes. At the bottom of the photograph there are the rectangular fields of the Improvement Period allotments of Newcastleton, above which there are the curvilinear banks of the pre-improvement fields which run up to and over-ride a curving medieval intake dyke. The hilltop is marked by the ramparts of a prehistoric fort.

FOREWORD

The focus of this book is on the character of Scotland's landscape. Arguably no other part of the natural heritage commands such public acclaim, or so strongly contributes to the image of Scotland internationally.

It is Scotland's scenery that underpins our tourism industry. Our distinctive landscapes are the most important reason for visitors coming to Scotland. However, we must always remember that people live and work in our countryside. For them in particular Scotland's landscape is a resource to be utilised. Landscape change is part of the nature of the countryside. This is especially true given the Government's strong support for rural communities and the rural economy, encouraging a dynamic and changing rural scene.

We recognise, however, that change must be handled sensitively and, wherever possible, in ways that enhance the landscape. Not only does this make good economic sense, it helps to minimise contention over change. The landscapes that we value today result from changes in the past. The landscapes that future generations value will have been influenced by landscape changes that we instigate today.

The Government have made a particular undertaking to protect and enhance Scotland's scenery. In 1997 I asked Scottish Natural Heritage (SNH) for advice in a number of areas, all of high importance and related to the care of our finest landscapes. For example, we are convinced of the case for the introduction of National Parks in Scotland. We believe that a National Park is the solution to the many years of debate about the best way to plan and manage Loch Lomond and the Trossachs. A National Park might also be an appropriate way forward in the Cairngorms, and possibly in a small number of other areas of great natural heritage importance.

I see National Parks as striking a balance between caring for areas of important natural heritage, including their natural beauty and biodiversity, and creating opportunities for local people to take care of their own social and economic development. The role of National Parks will be to foster sustainable development. I asked SNH to consult widely and to advise us on the form of National Park most appropriate to meet these ends with a view to legislation by the Scottish Parliament in due course.

The Government is also committed to the concept of a national landscape designation. I asked SNH to look again at the National Scenic Area (NSA) designation in the light of the move to create National Parks. This is a complex subject and will take some time to resolve, but I would expect it to be an area that the Scottish Parliament will wish to consider at an early stage.

If we are to have a sensible debate on these issues, there is a need for a sound factual background to inform the discussion. The Scottish Office commissions research from the Macaulay Land Use Research Institute on the development of quantitative techniques for producing spatial measures of landscape. This research is important in providing a framework for the analysis of the positive and the negative impacts on our landscape that may result from land use change, and from all facets of rural development including afforestation.

Every landscape is important to someone. It is with this in mind that I am delighted that the first full assessment has been made of the landscape character of Scotland. SNH has undertaken this across the whole of Scotland in collaboration with others, most notably the local authorities. The Landscape Character Assessment project has established a major inventory and database of enormous interest. It will provide a valuable input to policy-making and will allow landscape change to be guided more sensitively.

We have a managed, working countryside that needs to change to survive. The key is to steer that change sensitively and wisely, both for the people of Scotland and for the visitors to our country. I, therefore, welcome this book as a very useful contribution to a fascinating debate which must continue to attract and exercise us all in the years ahead.

John Sewel
Minister for Agriculture, the
Environment and Fisheries,
The Scottish Office
October 1998

LIST OF CONTRIBUTORS

S. Bell, Forestry Commission, 231 Corstorphine Road, Edinburgh EH12 7AT

S. Bennett, Environmental & Infrastructure Council Offices, Dumfries & Galloway Council, Engusu Street, Dumfries DG1 2PP

A. Bennett, Macaulay Land Use Research Institute, Craigiebuckler, Aberdeen AB15 8QH

R.V. Birnie, Macaulay Land Use Research Institute, Craigiebuckler, Aberdeen AB15 8QH

I.D. Bishop, Centre for GIS and Modelling, University of Melbourne, Parkville, Victoria, Melbourne 3052, Australia

D. Bryden, Tourism & Environmental Initiative, c/o Highlands & Islands Enterprise, Bridge House, 20 Bridge Street, Inverness IV1 1QR

N. Buchan, Scottish Natural Heritage, 2 Anderson Place, Edinburgh EH6 5NP

J.M. Bullen, Welsh Institute of Rural Studies, University of Wales, Llanbadarn Campus, Aberystwyth, Ceredigion SY23 3AL

L. Campbell, Scottish Natural Heritage, Wynne Edwards House, 17 Rubislaw Terrace, Aberdeen AB1 1XE

S. Carver, School of Geography, University of Leeds, Leeds LS2 9JT

P. Dixon, Royal Commission on Ancient and Historic Monuments of Scotland, John Sinclair House, 16 Bernard Terrace, Edinburgh EH8 9NX

W. Dramstad, The Norwegian Institute for Land Inventory, Raveien 9, PO Box 115, N-1430 Ås, Norway

H. Dufourmont, Vlaamse Landmaatschappij, Gulden Vlieslaan 72, B-1060 Brussels, Belgium

L. Dyson Bruce, Royal Commission on Ancient and Historic Monuments of Scotland, John Sinclair House, 16 Bernard Terrace, Edinburgh EH8 9NX

M. Fairbairn, Ordnance Survey, Romsey Road, Southampton SO16 4GU

S. Fritz, School of Geography, University of Leeds, Leeds LS2 9JT

G. Fry, NINA, PO Box 736, Sentrum, N-0105 Oslo, Norway

H. Gulinck, Laboratory for Forest, Nature and Landscape Research, Katholieke Universiteit Leuven, Vital Decosterstraat 102, B-3000 Leuven, Belgium

R. Gustavsson, Department of Landscape Planning Alnarp, Swedish University of Agricultural Sciences, PO Box 58, S-230 53 Alnarp, Sweden

D. Habron, SEPA East Region, Clearwater House, Heriot-Watt Research Park, Avenue North, Riccarton, Edinburgh EH14 4AP

C.M. Hägerhäll, Department of Landscape Planning Alnarp, Swedish University of Agricultural Sciences, PO Box 58, S-230 53 Alnarp, Sweden

R. Hingley, Historic Scotland, Longmore House, Salisbury Place, Edinburgh EH9 1SH

R. Hughes, Scottish Natural Heritage, 2 Anderson Place, Edinburgh EH6 5NP

A.C. Jones, The Geoinformation Group, 307 Cambridge Scheme Park, Cambridge CB4 4ZD

A. Law, Macaulay Land Use Research Institute, Craigiebuckler, Aberdeen AB15 8QH

S. Leadbeater, Macaulay Land Use Research Institute, Craigiebuckler, Aberdeen AB15 8QH

F. Lee Scottish Natural Heritage, Caspian House, Mariner Court, Clydebank Business Park, Clydebank G81 2NR

A. Lumb, Scottish Natural Heritage, Battleby, Redgorton, Perth PH1 3EW

L. MacInnes, Historic Scotland, Longmore House, Salisbury Place, Edinburgh EH9 1SH

E.C. Mackey, Scottish Natural Heritage, 2 Anderson Place, Edinburgh EH6 5NP

T.J. Maxwell, Macaulay Land Use Research Institute, Craigiebuckler, Aberdeen AB15 8QH

D.R. Miller, Macaulay Land Use Research Institute, Craigiebuckler, Aberdeen AB15 8QH

P. Minto, Countryside Council for Wales, Ffordd Penrhos, Bangor, Gwynedd LL57 2LQ

J. Moir, School of Town & Regional Planning, University of Dundee, 13 Perth Road, Dundee DD1 4HT

I. Nicol, Aberdeen City Council, Planning & Strategic Development, St Nicholas House, Broad Street, Aberdeen AB10 1BW

O. Puschmann, The Norwegian Institute for Land Inventory, Raveien 9, PO Box 115, N-1430 Ås, Norway

D. Rice, School of Town & Regional Planning, University of Dundee, 13 Perth Road, Dundee DD1 4HT

N. Shepherd, Forestry Authority, 231 Corstorphine Road, Edinburgh EH12 7AR

T.C. Smout, Chesterhill, Shore Road, Anstruther, Fife KY10 3DZ

I. Stas, Laboratory for Forest, Nature and Landscape Research, Katholieke Universiteit Leuven, Vital Decosterstraat 102, B-3000 Leuven, Belgium

J. Stevenson, Royal Commission on Ancient and Historic Monuments of Scotland, John Sinclair House, 16 Bernard Terrace, Edinburgh EH8 9NX

F. Thin, Scottish Natural Heritage, 27 Ardconnel Terrace, Inverness IV2 3AE

D. Tyldesley, David Tyldesley & Associates, Sherwood House, 144 Annesley Road, Hucknall, Nottingham NG15 7DD

M.B. Usher, Scottish Natural Heritage, 2 Anderson Place, Edinburgh EH6 5NP

A. Watt, School of Town & Regional Planning, University of Dundee, 13 Perth Road, Dundee DD1 4HT

C. Withers, Department of Geography, University of Edinburgh, Drummond Street, Edinburgh EH8 9XP

PART ONE

LANDSCAPE CHARACTER

1   THE LANDSCAPE CHARACTER ASSESSMENT OF SCOTLAND

Rebecca Hughes and Nigel Buchan

Summary

1.   SNH employs landscape character assessment (LCA) in order to meet part of its statutory remit.   Landscape character forms a fundamental part of the natural heritage of Scotland. LCA helps to increase the general awareness and understanding of this.   It is through the conservation, restoration or enhancement of landscape character that the diversity of Scotland's landscapes can be maintained for future generations to enjoy.

2.   LCA is now widely recognised as a structured, repeatable and defensible technique.   It is both an integrated and iterative approach, in which an understanding of all the physical, ecological and cultural processes impinging upon the landscape are analysed. The critical part of an assessment involves identifying the key characteristics of the landscape and the way in which these combine to form areas of consistent and recognisable landscape character.   LCA reports also contain guidelines which identify what effects development and other land use change may have upon landscape character in future.

3.   SNH initiated the LCA programme in order to establish an inventory of the landscapes of Scotland.   This will provide the context for dealing with a broad range of casework, assist SNH and other partners to influence both development plan policies and other land use strategies, and help to inform national policy.

4.   The SNH programme is based upon guidance originally developed by the Countryside Commission. A standard brief was used but, because one of the aims of the programme was to involve local authorities and other SNH partners from the outset, individual studies needed to reflect the aspirations of each project steering group.   LCA studies within the programme are consistent in the way they assess landscape character.   Most of the differences between studies are restricted to the approach to the development of landscape guidelines.

5.   The LCA programme is attracting widespread interest and it is encouraging to see that the descriptions and recommendations proposed in the reports are already being applied throughout Scotland. SNH promoted the LCA programme through the publication of individual LCA reports.   Other supporting publications will be distributed widely.

1.1 Introduction

This chapter examines the landscape remit of Scottish Natural Heritage (SNH), its approach to landscape character assessment (LCA) and the objectives, methodology and outputs of the national LCA programme. The final part of the paper deals with the potential applications of LCA.

The landscape remit of SNH is set out in its founding legislation. *The Natural Heritage (Scotland) Act 1991* states that the aims of SNH shall be to " secure the conservation and enhancement of, and to foster understanding and facilitate the enjoyment of, the natural heritage of Scotland". The Act goes on to define the natural heritage as including " the flora and fauna of Scotland, its geological and physiographic features, its natural beauty and amenity".

'Natural beauty' is understood by SNH to mean beauty that is not exclusively created by people (Plate 1), as distinct from the beauty of architecture or other works of art. Natural beauty, however, does not exclude beauty which arises as a result of, or despite, human activity. For example a reservoir may have considerable natural beauty, even though it is man-made, and the beauty is incidental to its purpose. Nor does natural beauty exclude the beauty of natural things when these have been deliberately placed or consciously arranged, as for example in a planned landscape. Thus, the concept of natural beauty includes the aesthetic qualities of landscape elements such as fields and field boundaries, shelterbelt and policy woodlands, plantations and even buildings where these elements are intrinsic to the character of a tract of landscape.

'Amenity' is a measure of the extent to which the natural heritage can be enjoyed by people. Places rich in wildlife, with varied land form, or of great natural beauty tend to have high amenity value, but location is also important. The natural heritage closer to where people live can have a high amenity value, simply because more people can regularly enjoy it. All landscapes may, therefore, be important to someone.

Throughout Scotland all landscapes have, to a greater or lesser extent been, and continue to be, modified by the activities of people as well as natural processes. Their natural beauty results, in varying degrees, from this interaction. The term 'natural beauty and amenity' is an important part of SNH's remit because it expresses something of the inherent variability of natural systems and the human role in them. It also reinforces our remit to 'facilitate the enjoyment of' the natural heritage, through the ability of landscape to provide people with first hand experience of the natural heritage.

Natural beauty is a term which embraces the landscape, but also the response of the individual to that landscape. It encompasses landform, land use, aesthetics and sensory qualities. An individual's perception of natural beauty will be influenced by cultural background, familiarity, experience and knowledge and is therefore infinitely variable.

It is through the landscape that SNH encompasses the human dimension of the appreciation of the natural heritage. By understanding the ways in which people relate to their environment, how they see it and how they behave within it, the study of landscape helps achieve a more holistic view of the natural heritage, rather than focusing on individual sites or specific components.

Through the process of LCA, SNH has chosen to concentrate on the more objective aspects of natural beauty. The term 'landscape character' is used to refer to the physical landscape resource and experience of that resource. It is a composite of these two components and of the natural and cultural processes which have formed and continue to shape the landscape. Landscape character is dynamic and continually in flux; it is also unique to a place, playing an important role in establishing local identity.

1.2 Landscape character in SNH

When SNH was established through the merger of the Countryside Commission for Scotland (CCS) and the Nature Conservancy Council for Scotland (NCCS) in 1992, it quickly became apparent that the data available on the landscape resource of Scotland was insignificant when compared to other aspects of the natural heritage, such as nature conservation. As a result, it was often difficult for SNH staff to assist and advise planning authorities, for example, on both individual development control cases and strategic planning issues. With the increasing emphasis placed upon the development plan it became important that SNH develop its understanding of the total landscape resource.

SNH therefore employed LCA as a way of meeting its remit to foster a better understanding and awareness of this part of the natural heritage and to increase the potential for enjoyment of natural beauty by the conservation, restoration or enhancement of landscape character in order that the great diversity of Scotland's landscapes may be maintained for future generations. In 1994 SNH embarked upon the nation-wide exercise known as the LCA programme. This was carried out with the co-operation of local planning authorities, but also involved other organisations such as the Forestry Authority, Forest Enterprise, Historic Scotland, enterprise companies, Scottish Office Agriculture Environment and Fisheries Department, and local groups such as the Shetland Amenity Trust, through consultations or the membership of project steering groups.

LCA involves the identification of the various component parts of a landscape and also an understanding of the ways in which these components interact. LCA is both an integrated and iterative approach, in which an understanding of all the physical, ecological and cultural processes impinging upon the landscape are analysed. The physical components of geology, geomorphology, soils, vegetation, ecology and climate give us an understanding of the evolution of the landscape and of the ongoing natural processes and relationships. An appreciation of the cultural influences on the landscape - such as patterns of settlement or cultivation - is necessary to understand how human intervention has changed the landscape over time.

The process of LCA explores the relationship between all these components and how humans experience them. People respond both consciously and sub-consciously to their surroundings. They can see shape, scale, variety, pattern, balance and edge, while their other senses enable them to experience a place through sounds, exposure, shelter, smells, vibration, human presence, movement and temperature.

The landscape can be assessed and classified through the process of LCA. By disassembling and analysing the component parts of a landscape, it is easier to understand what is important in the landscape, why it is important and how best to manage landscape change in the future. It helps with both managing adverse effects on the landscape as well as helping to direct methods of enhancement.

The LCA programme was initiated with the following five objectives:

- to establish an inventory of all the landscapes of Scotland;
- to provide the context for dealing with a broad range of casework, including development control and other land use changes;
- to provide information to assist SNH, local authorities and other partners in inputting to development plans and other land use strategies;
- to inform national policy, both within and outwith SNH; and
- to involve SNH's partners from the outset, and to encourage them to make maximum use of all products of the programme.

1.3 Method

In developing the programme, SNH carried out an appraisal of the methods that had been used previously in the United Kingdom and elsewhere in Europe and North America. Other countryside agencies were also developing landscape assessment programmes and an Inter-Agency Working Group was formed with the Countryside Commission, the Environment and Heritage Service in Northern Ireland and the Countryside Council for Wales, in order to share information and experience of LCA methodologies.

It was decided early on to base the SNH programme on the guidance developed by the Countryside Commission (1993) which, with some modification, it was felt would be largely applicable to Scotland. Earlier guidance, such as that produced by CCS (Land Use Consultants, 1991), had been directed more at designations and had attempted to evaluate landscapes in terms of their relative importance. The need to compile an inventory of landscape resources within a relatively short timescale required SNH to concentrate on the more straightforward aspects of LCA and to defer a consideration of landscape evaluation until after the inventory was completed.

Although the SNH programme has been based on the Countryside Commission guidance, the methodology has developed as work progressed, so as to take account of the wide diversity of Scottish landscapes. A standard brief was developed which was used for all studies forming part of the programme. Most studies were carried out at 1:50,000 scale, but there were exceptions where more detailed information was required. Although there is similarity in the way that landscape types and landscape character areas are defined, these reports reflect the diversity of uses to which studies have been put, as well as the aspirations of project steering groups. For example, the scale or detail of individual assessments was geared to the requirements of our project partners, and, in particular, the needs of local planning authorities.

LCA is an iterative process, involving familiarising, describing and understanding the landscape. It consists of a combination of background research, desk work (sieve mapping) and fieldwork. Background research includes an understanding of the historical, cultural and literary associations of a landscape, such as the perspective offered by artists, authors or poets. In all studies, the primary aim of the research has been to provide an understanding of the evolution of a landscape, the pressures that currently affect the landscape, and the changes likely to occur in future. In most cases it has been necessary to restrict the depth of research to reflect the broad scale of the studies.

Scotland is fortunate to have the detailed land use information provided by *Land Cover Scotland 1988* (Macaulay Land Use Research Institute, 1993). It has provided the basic land use information throughout the LCA programme. This has been supplemented by 1:24,000 scale black and white aerial photographs flown in 1988, the current Ordnance Survey 1:50,000 maps, and information about various designations, which have together formed the basis for the desk study.

Fieldwork consists of an initial period of familiarisation and the identification of field survey points, using the mapped information previously collected. An assessment of landscape character is made from each of these points using structured field survey sheets.

The critical part of a landscape assessment involves identifying the key characteristics of the landscape - what is important in the landscape and why? This involves the synthesis of the research, desk work and fieldwork, which is informed by professional judgement. Professional landscape architects with previous experience of LCA have been employed to carry out all studies within the programme, both for their knowledge of the subject but also to help maintain consistency between different studies.

From this synthesis it is possible to identify areas of consistent and recognisable landscape character. Where areas of similar landscape character occur, these can be reclassified into landscape character types. So, for example, in the Lochaber LCA (Environmental Resources Management, 1998) the same landscape type recurs on the island of Eigg and on Ardnamurchan, but these are two distinct landscape character areas.

Landscapes are dynamic. They are constantly evolving as a result of forces for change, such as afforestation, mineral working, housing in the countryside, or agricultural or other land use changes brought about by responses to the Common Agricultural Policy, developing technology or modern management systems. Each assessment identifies the key forces for change within each landscape type. By analysing the likely effect of future changes on the key characteristics of each landscape character type, it is possible to produce guidelines which can help direct future changes in the most appropriate direction and so help to maintain, reinforce or enhance the distinctive local landscape character. However, LCA guidelines avoid prescription, but instead outline the ways in which the key landscape characteristics may be affected by future changes, in order that appropriate decisions can be made that take these effects into account.

The progress of the LCA programme throughout Scotland has been influenced to some extent by the availability of partnership funding and other contributions, as well as other factors such as the timing of the development plan review process. For example, the Dumfries and Galloway LCA (Land Use Consultants, 1998) was timed to coincide with structure plan review, whereas the Aberdeen City LCA (Nicol *et al.*, 1996) was carried out jointly by staff of SNH and the local authority in order to contribute to the local plan review. Most studies have used the local authority boundary as their basis, although local authority reorganisation has meant that some studies will need to be presented in a different format in order to be of most use to the local authority.

Due to the need to tender each individual study competitively, a number of different consultants have contributed to the LCA programme. Other studies have been carried out by SNH and local authority staff.

Landscape character assessment is now widely recognised by the landscape and planning professions as a structured, repeatable technique. It is defensible, and has proven to be useful at various public local inquiries (for example in the Shieldaig inquiry, where the LCA was used to examine the effects of a proposed hydro development on the integrity of the National Scenic Area (NSA)).

1.4 Outputs of the LCA Programme

The results of the LCA programme will, in the first instance, be contained in 28 illustrated reports (listed in Table 1.1). The approximate spatial extent of each of these studies is shown in Figure 1.1. Each study in the programme involved local authorities and other SNH partners from the outset, and individual studies reflected the aspirations of each project steering group. In some cases this may have determined the scale or extent of the study, whereas in others it resulted in the detailed consideration of a particular force for change, such as afforestation or mineral development. Despite these different aims, a high degree of consistency has been maintained throughout the programme by the adoption of a standard brief which set out the basic methodology for each study.

Copies of all LCA reports are held in SNH offices and have been distributed to public libraries, members of project steering groups and other partners. Further copies are available from SNH Publications, Battleby, Perth.

Landscape character information is also held digitally in the form of a national Geographical Information System (GIS) dataset of landscape character types which is linked to a database of key characteristics for each of the 365 landscape character types identified. The database also identifies the predominant forces for change that affect each landscape character type. This digital information will provide greater ease of access by SNH staff to the basic LCA information, while reference to the LCA reports will remain necessary for a fuller understanding of the landscape character of any area. A map summarising the results of the LCA programme is provided in Plate 2.

SNH is currently working to identify and categorise the landscape character units and types by key characteristics so that the data can be reclassified in various ways and at different levels of resolution for a variety of purposes.

Since 1993, LCA methodology has evolved and developed, largely as a result of its widespread use. Recent discussions with the other countryside agencies have identified the need to produce updated guidance on all aspects of LCA and SNH is currently working with the Countryside Agency to develop this.

1.5 Applications of LCA

The LCA programme is attracting wide interest and it is encouraging to see that the descriptions and guidelines in the reports are being more widely applied as time goes on. The following applications are helping to reinforce the awareness and understanding of landscape character, both in the wider countryside and of the designated areas.

Locally, at the most detailed level, at the level of the individual development proposal or other land use change, the *Guidelines for Landscape and Visual Impact Assessment* (Landscape Institute and Institute of Environmental Assessment, 1995) recommend that LCA plays a key role in determining the landscape impact of any land use changes. The guidelines are now widely adopted by professional landscape architects and land use planners throughout the UK. Landscape character assessment is also useful in a more strategic way, as a tool that can be used at a number of different levels, to determine sensitivities to particular types of

Table 1.1. SNH Reviews of landscape character assessment. Each Review covers a local area, and includes maps to indicate the area covered. In a few instances there is overlap between two reports.

No.	Title	Date	Author(s)
19	Dunfermline District landscape assessment	1996	Tyldesley, David and Associates
37	National programme of landscape character assessment: Banff and Buchan	1994	Cobham Resource Consultants
71	Skye and Lochalsh landscape assessment	1996	Stanton, C.
75	Cairngorms landscape assessment	1996	Turnbull Jeffrey Partnership
77	The landscape of Kinross-shire	1995	Tyldesley, David and Associates
78	Landscape assessment of Argyll and the Firth of Clyde	1996	Environmental Resources Management
79	Mar Lodge landscape assessment	1996	Turnbull Jeffrey Partnership
80	Landscape character assessment of Aberdeen	1997	Nicol, I., Johnston, A. and Campbell, L.
90	Inner Moray Firth landscape character assessment	1997	Fletcher, S.
91	The Lothians landscape character assessment	1997	ASH Consulting Group
92	Western Isles landscape character assessment	1998	Richards, J.
93	A landscape assessment of the Shetland Isles	1998	Gillespies
94	Dumfries and Galloway landscape assessment	1998	Land Use Consultants
96	Clackmannanshire landscape character assessment	1998	ASH Consulting Group
97	Lochaber: landscape character assessment	1998	Environmental Resources Management
100	Orkney landscape character assessment	1998	Land Use Consultants
101	Moray and Nairn landscape assessment	1998	Turnbull Jeffrey Partnership
102	South and Central Aberdeenshire landscape character assessment	1998	Environmental Resources Management
103	Caithness and Sutherland landscape character assessment	1998	Stanton, C.
111	Ayrshire landscape assessment	1998	Land Use Consultants
112	The Borders landscape assessment	1999	ASH Consulting Group
113	Fife landscape character assessment	1999	Tyldesley, David and Associates
114	Inverness District landscape assessment	1999	Richards, J.
116	Glasgow and the Clyde Valley landscape assessment	1999	Land Use Consultants
119	Ross and Cromarty landscape character assessment	1999	Ferguson McIlveen
120	Ben Alder, Ardverikie and Creag Meagaidh landscape character assessment	1999	Landscape Group, Advisory Services, SNH
122	Tayside landscape character assessment	1999	Land Use Consultants
123	Central Region landscape character assessment	1999	ASH Consulting Group
124	Stirling to Grangemouth landscape character assessment	1999	Tyldesley, David and Associates
LLT	Loch Lomond and the Trossachs	-	To be completed

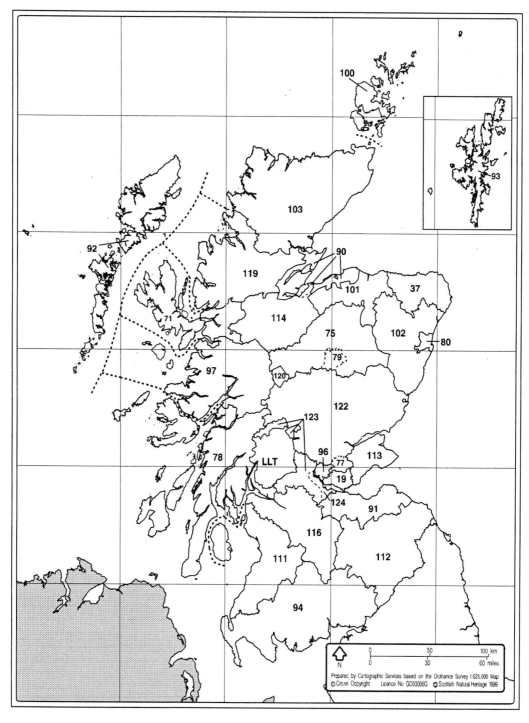

Figure 1.1. Map showing the spatial extent of each of the landscape character assessment studies. The numbers indicate the SNH Reviews as listed in Table 1.1.

landscape change. In these cases, LCA has been used to identify and analyse the key landscape characteristics and forces for change, so that the tolerance or capacity of the landscape to accept change can be identified.

Regionally, LCAs have been useful to local authorities in taking forward their structure plan review process. For example, the Central Region LCA (ASH Consulting Group, 1999) provided information in relation to the structure plan review and has also assisted in

- working towards sustainable development,
- informing strategic forestry policies, and
- producing a review of the local landscape designations (as well as the positive management and promotion of these).

Bennett (this volume) discusses how the LCA for Dumfries and Galloway has been used as a working paper in the structure plan review process.

Regional LCAs at 1:50,000 scale can also provide the framework for more detailed plans and strategies. For example, Stirling Council plan to take the Central Region LCA (ASH Consulting Group, 1999) and use it in combination with other information, such as the market pressures for development and ease of access to public transport, to identify areas of search. These will be subject to a detailed LCA at 1:10,000 scale. The purpose of the assessment will be to determine the capacity of the landscape to accept development, in order to assist the Council to identify potential locations in its draft housing/development strategy. The assessment will also be used to identify any landscape design requirements - such as the need to reinforce the existing landscape framework - where the landscape has the capacity to accept development.

In response to the large number of applications for open-cast coal mining in Clackmannanshire, the Council and SNH are working together to develop an open-cast coal strategy. This will involve interpretation of the 1:50,000 scale LCA, together with information about the supply and demand for coal and some further field work. It is intended that three categories of sensitivity to open-cast mining will be defined, namely

- 'restricted', where landscape character is likely to be irreversibly changed;
- 'constraint areas', where planning consent will be given only where the need for coal outweighs the significance of the landscape impacts; and
- 'controlled areas', where consent may be granted, subject to conditions that limit the impact on the landscape.

Other examples include a draft wind energy strategy for Orkney Islands Council (Orkney Islands Council, 1997) and landscape design guidance for forests and woodlands in Dumfries and Galloway (Environmental Resources Management, pers. comm.). The Dumfries and Galloway study will combine the guidelines provided by the regional LCA with Forestry Commission design guidance to provide practical advice for the integration of new woodland and forests. Guidance sheets, intended to help grant scheme applicants develop appropriate planting proposals, have been produced for each of the landscape types where forestry is an issue. Shepherd and Bell (this volume) describe this work in more detail.

SNH will shortly be carrying out a review of the use of LCA in development plans and strategies in order to establish best practice guidance, based on experience elsewhere throughout the UK.

Landscape character assessments are also being used as a basis for further work to develop management strategies. For example, at Mar Lodge, SNH and the National Trust for Scotland sponsored a detailed landscape assessment of the whole estate (Turnbull Jeffrey Partnership, 1996b). This used the wider Cairngorms LCA (Turnbull Jeffrey Partnership, 1996a) as its basis but was able to examine the landscape implications of land use change at the level of detail necessary to feed into the estate management plan. It allowed detailed recommendations to be made that took account of the need to maintain landscape diversity.

On Rum, which is a National Nature Reserve (NNR) owned by SNH, a Historic Land Use Assessment is being carried out which will identify the predominant historical interests in the landscape and will complement the proposed LCA. Together, these studies will ensure that these aspects of the natural and built heritage are taken into account within the management plan for the NNR. SNH is currently working with Historic Scotland to explore the potential for integrating the two methods in other parts of Scotland. Dixon *et al.* (this volume) discuss this in more detail.

At a national level, SNH will be using the national landscape character dataset in a number of ways. The Natural Heritage Zones (NHZ) programme for Scotland will provide a framework of 21 zones, within which objectives and priorities for the natural heritage will be set in a way that is responsive to both local needs and national priorities. It will provide a further link between designated sites and the wider countryside. The zones have been derived from SNH's work on biogeographical zonation (Usher and Balharry, 1996) and LCA. Information from the LCA programme is to be integrated throughout all stages and products of the NHZ programme, to ensure that this aspect of SNH's remit is more comprehensively understood. Thin (this volume) describes the aims and objectives of the NHZ programme in more detail.

As part of the designations review, being carried out by the Scottish Office in 1998, SNH has been asked to review the suite of NSAs and to make recommendations for any changes in the selection or extent of the current NSAs. SNH will be using the national dataset to examine the landscape characteristics of the NSAs to help to inform discussions.

A pilot study to examine the application of the national dataset to the potential for wind farm development is also planned in response to a Scottish Office request to identify potential areas in Scotland. The study will examine the opportunities for linking the key characteristics of the national dataset to the capability of the landscape to accept wind farm development.

Increasingly, LCA information is being cited in national policy guidance, such as Planning Advice Notes (PANs) and National Planning Policy Guidelines (NPPGs). The NPPG on the Natural Heritage (Scottish Office, 1999), for example, states "some planning authorities have commissioned assessments of landscape character. Such assessments can provide valuable local guidance on the capacity of the landscape to accommodate new development. The scale, siting and design of new development should take full account of the character of the local landscape".

1.6 Conclusions

Landscape character assessment has an important role in increasing the general awareness of all aspects of the natural heritage. To many, the terms 'landscape' or 'landscaping' conjure images of earth mounding or block and barrier planting that attempts to conceal developments that could have been sited or designed with more respect for the local landscape character in the first place. SNH has embarked on this programme of LCA to help to increase the general level of knowledge and understanding of the enormous variety, and unique character, of the landscapes of Scotland. Development and other land use changes must respect or build on this local distinctiveness if we are to avoid erosion of landscape character and the sort of suburbanisation that causes all places to seem the same.

SNH has promoted, and continues to promote, the LCA programme through the publication of the individual assessment reports and of a colour leaflet (Anon., 1998). Further publications will help to ensure that a wide audience is made aware of the work. For example, a summary report of the LCA programme, which will be similar in format to the publication *The Natural Heritage of Scotland - an Overview* (SNH, 1995), is intended to inform 'state of the environment' reporting. At a more local level, there are plans to produce leaflets publicising and explaining individual LCAs. The first of these, for Skye and Lochalsh, is currently being developed.

For the very first time an inventory of the landscape character of Scotland is available, enabling all those making decisions about land use change in Scotland - whether government, local authorities, other organisations or landowners - to do so with a more comprehensive understanding of the implications for the immensely varied and distinctive landscapes of Scotland.

References

Anon. (1998). *The Landscape Character of Scotland: the National Programme of Landscape Character Assessment.* Perth, Scottish Natural Heritage.

ASH Consulting Group (1999). *Central Region Landscape Assessment.* Scottish Natural Heritage Review No. 123

Countryside Commission (1993). *Landscape Assessment Guidance.* Cheltenham, Countryside Commission.

Environmental Resources Management (1998). *Lochaber: Landscape Character Assessment.* Scottish Natural Heritage Review No. 97.

Land Use Consultants (1991). *Landscape Assessment Principles and Practice.* Perth, Countryside Commission for Scotland.

Land Use Consultants (1998). *Dumfries and Galloway Landscape Assessment.* Scottish Natural Heritage Review No. 94.

Landscape Institute and Institute of Environmental Assessment (1995). *Guidelines for Landscape and Visual Impact Assessment.* London, E&FN Spon.

Macaulay Land Use Research Institute (1993). *The Land Cover of Scotland 1998.* Aberdeen, Macaulay Land Use Research Institute.

Nicol, I., Johnston, A. and Campbell, L. (1996). *Landscape Character Assessment of Aberdeen.* Scottish Natural Heritage Review No. 80.

Orkney Islands Council (1997). Landscape Assessment, Preferred Areas for Wind Turbine Development. Unpublished report, Orkney Islands Council.

Scottish Natural Heritage (1995). *The Natural Heritage of Scotland: an Overview.* Perth, Scottish Natural Heritage.

Scottish Office (1999). *Natural Planning Policy Guideline, NPPG14: Natural Heritage.* Edinburgh, The Scottish Office.

Turnbull Jeffrey Partnership (1996a). *Cairngorms Landscape Assessment.* Scottish Natural Heritage Review No. 75.

Turnbull Jeffrey Partnership (1996b). *Mar Lodge Estate: Landscape Assessment.* Scottish Natural Heritage Review No. 79.

Usher, M.B. and Balharry, D. (1996). *Biogeographical Zonation of Scotland.* Perth, Scottish Natural Heritage.

2 THE USE OF LANDSCAPE CHARACTER IN DEVELOPMENT PLANNING: TWO CASE STUDIES

S.P. Bennett, L. Campbell and I. Nicol

Summary

1.   Case studies of the use of landscape character in Dumfries and Galloway and the City of Aberdeen are given. Both studies formed part of the national coverage of Landscape Character Assessment (LCA) in Scotland.

2.   The Dumfries and Galloway assessment has informed various aspects of the Structure Plan review, including the Wind Energy Strategy, Landscape Design Guidance for Forests and Woodlands, Forestry Frameworks and proposed Renewable Energy, Forestry and Minerals Subject Local Plans, as well as influencing general policy development and providing a framework for the Regional Scenic Area review.

3.   The Dumfries and Galloway assessment is also informing the ongoing Local Plan reviews and is used in assessing planning applications and Woodland Grant Scheme applications.

4.   The Aberdeen City assessment is currently used in planning casework and it is hoped to incorporate it into the Local Plan.

5.   Future use of the Aberdeen assessment is likely to include contributing to a landscape designation review, housing capacity study, and a landscape strategy for the city.

2.1 Introduction

This chapter presents two case studies of the use of landscape character assessments in very different parts of Scotland. The first case study focuses on Dumfries and Galloway, a large rural area in the south west of Scotland where forestry and wind energy raise significant landscape issues. In contrast, the second case study explores the largely urban situation of the City of Aberdeen, where high rates of development, the need to conserve the setting of the city, and the lack of clarity about green belt designation around the city are the important issues.

2.2 Dumfries and Galloway

Dumfries and Galloway is a relatively sparsely populated area, with an ageing and now declining population concentrated towards the coast and the main river valleys. The principal land uses are agriculture, in the form of dairy farming and stock rearing, and forestry. Tourism supports around 10% of all employment.

   The landscape varies from the rounded hills of the Southern Uplands and the more rugged peaks of central Galloway to wide river valleys, estuaries of international importance for their wildlife, and rocky peninsulas. Development pressures are not intense because of the area's peripherality, but the area is attractive to both the forestry and windfarm industries, as well as being subject to the ubiquitous pressures for expansion to the main settlements, infrastructure improvements and new buildings in the countryside, along with tourist developments.

2.2.1 The Landscape Assessment

In the autumn of 1993, the then Regional Council announced a review of the Structure Plan for Dumfries and Galloway, and agreed that this should be informed by a regional landscape assessment. The Council formed a partnership with three other organizations: Scottish Natural Heritage (SNH), who were similarly interested in taking forward such regional scale assessments, the Forestry Commission (FC), as the agency involved with one of our most significant forms of land use change, and Dumfries and Galloway Enterprise. In 1994 Land Use Consultants were commissioned to take forward what was to be Scotland's first regional scale landscape assessment (Land Use Consultants, 1998).

The Report comprises three sections. The first section reviews the forces which have shaped the landscape of the region and highlights key features. The second section summarises the various changes happening in the landscape through changing agricultural practices, forest and woodland planting and management, and development pressures. Relevant development issues include urban expansion on the edges of the main settlements, new agricultural buildings and dwellings in the countryside, road improvements, linear power supplies such as new gas pipelines, and interest in windfarms and tourist developments. The third section divides the region into four Regional Character Areas comprising the Rhins and Machars, Galloway Uplands, West Southern Uplands and Dumfries Coastlands. These are divided into a series of 27 repeating landscape character types, including a range of coastal, lowland, valley and upland landscape types. The latter are split into a series of predominantly forested or unforested landscapes (see Plate 3). The individual landscape character types are described, key issues are highlighted, and an overall strategy is proposed for each character type, followed by a series of guidelines on the principal relevant forces for change.

2.2.2 Applications

The assessment has already informed various aspects of the Finalised Structure Plan (Dumfries and Galloway Council, 1998a). It has influenced the formulation of general policies on the landscape impacts of development, and has provided a framework for reviewing the Regional Scenic Designations. It is also being used to develop more detailed policies and guidance on specific forms of development and land use change through

- the Dumfries and Galloway Wind Energy Strategy,
- the proposed Renewable Energy Subject Local Plan,
- Landscape Design Guidance for Forests and Woodlands in Dumfries and Galloway,
- Local Forestry Frameworks,
- the proposed Forestry Subject Local Plan, and
- the proposed Minerals Subject Local Plan.

The assessment is also informing the ongoing Local Plan reviews, and is used in evaluating individual development proposals and Woodland Grant Scheme applications.

The Landscape Design Guidance for Forests and Woodlands and the two local Forestry Frameworks are both partnership projects with SNH and FC, and are discussed by Bell and Shepherd (this volume). This chapter therefore concentrates on the general landscape policies of the Structure Plan, the Regional Scenic Area review (Dumfries and Galloway Council, 1997) and the Windfarm Strategy (Dumfries and Galloway Council, 1998b).

2.2.3 General policy development in the Structure Plan

The Structure Plan chapter on 'Caring for the Environment' includes Policy E3 which states that 'When assessing development proposals likely to have a significant impact on the landscape the Council will take into account the guidance set out in the Landscape Assessment'. The overall strategies for each character type are summarised in a map indicating landscapes whose character is worthy of conservation, is worthy of conservation but would benefit from enhancing, or would benefit from enhancement or restoration (Figure 2.1). This map provides a framework for targeting resources for landscape conservation and enhancement. It also offers an initial indication of those landscapes which may be sensitive to change, versus those whose character is less well developed or is being degraded, and may be more capable of accommodating change (though it is important to refer back to the characteristics of the landscape type). Many other policies relating to specific forms of development and land-use change also make reference to landscape issues.

2.2.4 Regional Scenic Area review

Only 3 percent of Dumfries and Galloway has National Scenic Area status, so local scenic designations remain an important mechanism for landscape conservation. The 1984 Structure Plan identified a series of Areas of Regional Scenic Significance. These took account of previously defined Areas of Great Landscape Value, established in response to the requirement of the Scottish Development Department. However, there was no systematic basis for evaluation, and individual planning officers tended to use different criteria to select areas and define boundaries.

The landscape assessment formed the basis of a systematic review of the designated areas in the form of a background paper (Dumfries and Galloway Council, 1997). The review set out to identify good examples and attractive combinations of scenically valued landscape types which form recognisable and comprehensible geographic units. A variety of factors was taken into account including existing and past designations, recognisable identity and local distinctiveness, typicality and uniqueness, wild land values, visual prominence, accessibility and sensitivity to change. Boundaries were related to defined attributes such as visual envelopes, topographic units or landscape character type boundaries. As a result ten Regional Scenic Areas have been defined (Figure 2.2). The developing policy framework gives particular priority to landscape conservation in these areas.

2.2.5 Wind Energy Strategy

The Structure Plan is supported by a Wind Energy Strategy (Dumfries and Galloway Council, 1998b). This classifies areas of high wind speed, in which windfarms are potentially feasible, into sensitive, intermediate and potential areas in terms of their

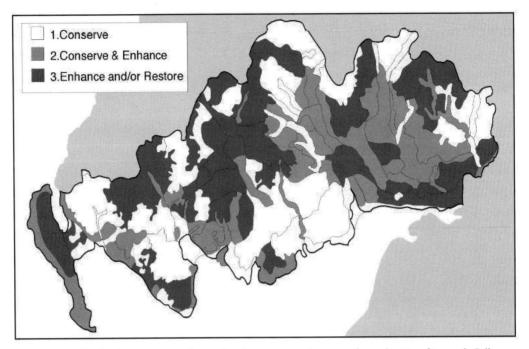

Figure 2.1. The landscape conservation and enhancement policy map from the Dumfries and Galloway Finalised Structure Plan.

sensitivity to wind farm developments (Figure 2.3). Structure Plan Policy S23 provides a presumption against wind farm developments in sensitive areas, whilst indicating that proposals in intermediate and potential search areas may be considered favourably where they meet the guidance in the Strategy.

The Strategy considers a range of issues including impacts on nature conservation, cultural heritage, land use, tourism, recreation, people, roads and settlements. Landscape and visual impacts are given particular weight because they are usually high on the list of potential concerns. They are addressed in three separate ways.

First, the landscape assessment examines the sensitivity of individual landscape types to windfarm developments and ascribes each to one of three categories in terms of their potential as search areas for windfarm sites. This advice has since been reviewed in the light of the increasing size and number of turbines in more recent proposals, our experience of the recently-constructed Windy Standards windfarm, and local landscape preferences revealed in the consultation process. Together, this information has been used to allocate individual landscape types to the sensitive, intermediate and potential categories used in the Strategy, except where other sensitivities raise them to a more sensitive category. The specific sensitivities of each landscape type are then outlined in the text.

Second, National and Regional Scenic Areas are automatically allocated to the sensitive category because they represent those parts of the region which are particularly valued for their existing character.

Third, the Strategy also provides detailed advice on the various factors which influence landscape and visual impacts. This includes a discussion of cumulative impacts and offer guidance on the siting and design of individual proposals.

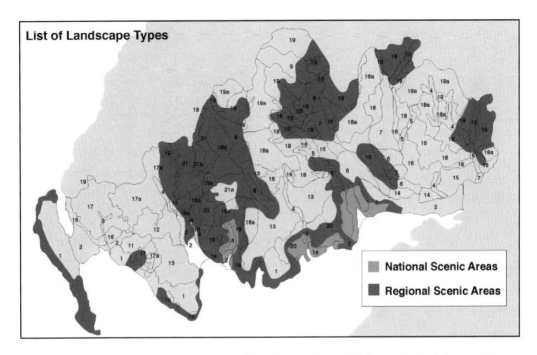

Figure 2.2. National and Regional Scenic Areas from the Dumfries and Galloway Finalised Structure Plan.

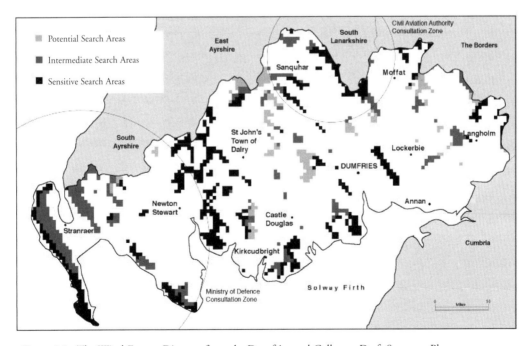

Figure 2.3. The Wind Energy Diagram from the Dumfries and Galloway Draft Structure Plan.

The Council intends to develop the findings of the Wind Energy Strategy further as part of the proposed Renewable Energy Subject Local Plan.

2.3 Aberdeen City

Aberdeen is the third largest city in Scotland, with a population of just under 220,000 in an area of 185 km^2 (72 square miles). It is a coastal and harbour town, largely situated between the valleys of the Rivers Dee and Don, and serves a large hinterland in the north-east of the country. The distinctive background provided by the sea, river valleys, and ring of low hills to the west has had a marked influence on the development of the city. The dominant use of the local stone for many of its buildings has led to Aberdeen being known as 'the Granite City'. This, together with its location and setting, has created a strong sense of identity.

Due to its pivotal role in North Sea oil and gas exploration during the 1970s and 1980s, Aberdeen gained the status of 'Europe's Oil Capital'. During this time, development planners tended to focus on the growth of the urban area in order to cope with the rapid expansion of the city. Relatively scant attention was given to rural areas, other than establishing a Green Belt to limit development. Under the terms of Government legislation (Anon., 1985) one of the main reasons for the designation of Green Belts was to protect the landscape setting of towns and cities. Despite this, there was little or no attempt to analyse the landscape resource of the city at that time.

Aberdeen has remained relatively prosperous and this is reflected in overall development rates, with around 1,200 houses built every year. However, the Council and the business community wish to avoid over-reliance on oil. The encouragement of other industrial and economic developments has therefore been seen as important. Consequently, the built-up area of the city has expanded by about 10 km^2 (4 square miles) over the last twenty years. Some of this development was built on 'greenfield' sites, with varying levels of impact on the landscape setting of the city, despite the protection that the Green Belt was intended to provide. The lack of clarity in the initial reasons for the Green Belt designation around Aberdeen can impede current planning decisions and, occasionally, open them to challenge by developers.

2.3.1 The Landscape Character Assessment

This combination of circumstances led several planners in Aberdeen City Council to realise that the distinctive setting of the city was important; that the high rates of development could adversely affect it; and that there was a lack of knowledge about the landscape of the area, particularly the Green Belt and the reasons for its designation. The planners believed that a landscape character assessment would be useful in clarifying these issues. It was also hoped that it would provide a basis for policy review, particularly of the Green Belt.

Accordingly, in 1994 a small team of planners from the City Council carried out a pilot study using the Countryside Commission methodology (Countryside Commission, 1993). However, the inexperience of the survey team in dealing with landscape character assessment, and the problems of using methods developed for larger regionally-based studies and not for smaller urban areas, soon became apparent. Therefore, the City Council turned to SNH for assistance. Under its national programme a new study was begun in 1995, which involved a steering group of staff drawn from the Council and SNH. The desk survey, fieldwork, classification and drafting of the report were carried out by one member of staff from each organisation (Nicol *et al.*, 1996).

The report identified 27 individual character areas within five landscape character types (Plate 4). The assessment was carried out at the more detailed scale of 1:25,000 because of the relatively small size of the study area. This also provided the level of detail that was needed to meet the requirements for planning policy and development control purposes.

Carrying out the assessment with an in-house team had both advantages and disadvantages. The report took a lot of staff time to produce, but thorough fieldwork was more easily accomplished because of the team's detailed knowledge about the city. The presence of development control planners on the steering group meant that their concerns from a user's point of view, particularly on the detailed wording of the report, could be addressed from an early stage.

A member of the council archaeology unit provided information describing the evolution of the city. The archaeologist and the survey team worked together to ensure that the account of Aberdeen's history was focused on relevant landscape change. In turn, including an archaeologist on the team raised awareness of the historical aspects of the landscape so that it was adequately considered during the study process.

In addition to the standard Countryside Commission methodology, the Aberdeen study included a chapter on visual aspects. This attempted to identify the countryside features that are most visible from the city and which, therefore, would most affect the setting of the city if they were built-upon. Selected viewpoints along the main routes into the city were used to assess the degree of visibility of features and to describe wider visual aspects. The study also tried to analyse what made the approaches to the city distinctive, for example the avenues of trees and church spires, and which notable city landmarks were visible from the main approach routes. In addition, the level of visibility of an area from major viewpoints was used as one of the criteria in assessing the sensitivity to change of the various landscape character areas.

2.3.2 Current use

Both the Council and SNH have used the landscape character assessment in the evaluation of planning applications. These have included proposals for landfill operations, coastal protection, golf course construction, woodland grant schemes and major housing developments. The landscape character assessment was recently put under scrutiny at a local Public Inquiry which concerned an appeal to the Secretary of State on the refusal of a planning application for disposal of waste by land-raising rather than land-filling at a site near an approach route to the city. The landscape character assessment appeared to stand up well to cross-examination.

It is too early to see if the landscape character assessment has made a difference 'on the ground'. However, it has definitely been useful in establishing and clarifying landscape issues in many cases. Its status, as a published document produced jointly by the Council and SNH, has aided its acceptance by non-specialist staff both inside and outside these organisations.

The review of the Aberdeen Local Plan (Aberdeen City Council, 1998) provided an opportunity to incorporate the landscape character assessment into the development plan. The draft plan recognised the importance of the city landscape and its setting. However, it seemed that several areas of uncertainty remained and the potential contribution of the landscape character assessment had not been fully realised. For example, a system of landscape designation is proposed that does not relate clearly to the assessment. This was perhaps due to confusion between landscape assessment, which involves relatively objective identification and classification, and landscape evaluation, which applies values to the landscape. The assessment tried to make it clear that it provided information on only one of a range of factors which need to be considered before reaching a balanced view on development issues; it did not consider some landscapes to be more valuable than others. It is hoped that the finalised Local Plan will remedy these misunderstandings and include the landscape character assessment more fully and appropriately.

Not using the assessment to its full potential may be symptomatic of the fact that some planners often confuse the definition of landscape with the process of landscaping. This can result, for example, in requesting the use of planting or mounding to hide developments rather than assessing whether the developments correspond adequately to the surrounding landscape pattern or not. The assessment should be able to overcome this false impression by helping to raise awareness about the intrinsic nature of landscape and its importance. Some of the planners on the steering group, overseeing preparation of the landscape character assessment, found it a beneficial process. It made them look harder at the landscape and give greater consideration to the context of developments. One way of clarifying these issues could be by developing more formal training opportunities, such as local Continuing Professional Development seminars for staff. Ideally, these seminars should involve staff from SNH and other agencies.

2.3.3 The future

The Green Belt around Aberdeen is designed to protect the landscape setting of the city. There is an opinion that its constraint on development should be modified or made more flexible, but any proposed changes would need to be tempered by objectives relating to sustainability. In order to respond to the various issues involved in reassessing the function and boundaries of the Green Belt, a multi-disciplinary team will be convened to look at the wider environmental, as well as social and economic questions raised by city developments. The landscape character assessment will provide both the essential information on the landscape resource and a base from which to develop sound criteria for designation.

A significant proportion of the pressure for development in and around the city comes from housing. Recent government targets have been set for 'brownfield' development using sites within the town, but this is unlikely to meet demand. The need to avoid 'town cramming', and the potential threat to some distinctive urban landscape features, may require consideration of some 'greenfield' development. There are signs that the combined effects of this targeting, and the constraints exerted by the Green Belt, are causing some adverse impacts on urban areas. Examples of this can include higher housing densities, garden splitting to provide new sites, smaller gardens and open spaces, and larger developers taking over from small builders. It could be argued that around 85 percent of the total population of the city area, that lives in the urban parts, is experiencing some loss in local amenity in order to protect the landscape setting of the city and the more rural areas where only some 15 percent of the population lives.

It is intended to use the landscape character assessment, including its visual analysis section, as the basis for analysing the capacity of the landscape to accommodate additional development. This should help to ensure that any expansion into the rural area around the city is carried out in a more balanced manner.

A further likely step is the preparation of a Landscape Strategy for the city. This will identify, at a more detailed scale, the areas and landscape features which contribute most to the setting of the city. The strategy will then consider appropriate steps to safeguard these areas and features; put forward environmental project work for conservation, enhancement, and restoration; establish a programme of management for key areas of open space and important landscape features; and produce landscape design guidelines for new developments.

2.4 Conclusions

There are considerable differences in the scale, scope and characteristics of the two studies described in this chapter. The study area for Dumfries and Galloway is over thirty times as large as that for Aberdeen. The former concerns a largely rural environment while the latter is predominantly urban. Both areas are removed from the main economic and population centres of the Central Belt of Scotland, but the differences between them have resulted in contrasting pressures for change. There are also differences in the ways in which each of the studies was prepared, although both adhered to the Countryside Commission guidance (Countryside Commission 1993).

Both landscape character assessments were experimental to some extent: Dumfries and Galloway by virtue of it being one of the first in Scotland and Aberdeen by being done in-house, at a more detailed scale, and including an attempt at visual analysis of the city. The adaptations used for Aberdeen City would be unlikely to work at the scale of Dumfries and Galloway. In contrast with the 'in-house' method used in Aberdeen, the Dumfries and Galloway landscape character assessment was notable for the extent of collaboration with the Forestry Commission. This was important given the scale of forestry development and redevelopment in the region.

At a basic level, both landscape character assessments are being used for similar purposes, albeit at different scales and with slightly different foci. Both assessments offer a clear and accessible framework for considering landscape issues, and a starting point for preparing more detailed strategies, guidance and policies to integrate landscape character more fully into the development planning process.

The Dumfries and Galloway study has been available for longer and has been more thoroughly incorporated into statutory plans. The 1:50,000 scale at which it was undertaken makes it particularly relevant to strategic planning for developments such as forestry and wind farms. The landscape characterisations and design advice have also proved useful in assessing some planning applications and mitigation measures for other forms of development, and in directing certain landscape management initiatives. The 1:25,000 scale of the Aberdeen study is more suited to planning for urban expansion and housing, and identifying specific management and enhancement projects. Aberdeen City Council is only beginning to use the assessment in its local plan, but having the assessment available has helped, to some extent at least, to raise awareness of landscape issues.

Undertaking the landscape character assessments has been both interesting and valuable. As a result, more is understood about the landscapes of areas as different as Dumfries and

Galloway and the City of Aberdeen. In addition to making a contribution to development control casework, both studies are being used as the basis for further work so as to ensure that future developments do not compromise the distinctive landscape character of the areas.

References

Aberdeen City Council 1998. *Aberdeen City Local Plan Consultative Draft.* Aberdeen, Aberdeen City Council.

Anonymous 1985. *Circular 24/1985: Development in the Green Belt and Countryside.* Edinburgh, Scottish Development Department.

Countryside Commission 1993. *Landscape Assessment Guidance CCP423.* Cheltenham, Countryside Commission.

Dumfries and Galloway Council 1997. *Identification of Regional Scenic Areas.* Dumfries and Galloway Council Technical Paper No. 6.

Dumfries and Galloway Council 1998a. *Finalised Structure Plan.* Dumfries, Dumfries and Galloway Council.

Dumfries and Galloway Council 1998b. *Preparation of Wind Energy Diagram.* Dumfries and Galloway Council Technical Paper No. 5.

Land Use Consultants 1998. *Dumfries and Galloway Landscape Assessment.* Scottish Natural Heritage Review No. 94.

Nicol, I., Johnston, A. and Campbell, L. 1996. *Landscape Character Assessment of Aberdeen.* Scottish Natural Heritage Review No 80.

3 LANDSCAPE ASSESSMENT IN THE NATURAL HERITAGE ZONES PROGRAMME

Frances Thin

Summary

1. In contributing to the debate on sustainable development, landscape is an important and valuable concept because, by its very nature, it is integrated and people focused.

2. Scottish Natural Heritage has a complex remit spanning nature conservation, landscape, recreation and access, environmental education and enjoyment. The Natural Heritage Zones Programme aims to provide a strategy that will integrate thinking across this remit and inform the actions of individuals and organisations which influence the natural heritage.

3. The impetus for the programme is to gain a better understanding of, recognition of , and commitment for the natural heritage and its social and economic role at both local and national levels.

4. Landscape is drawn into the programme primarily through the process and products of landscape character assessment and built into a series of inter-related outputs which are designed for different purposes and target audiences. Initial outputs from the programme are scheduled to be completed by the end of 2000. The thinking behind the Natural Heritage Zones Programme and its application, however, are based on a 25 year timescale.

5. The Natural Heritage Zones Programme will provide a rigorous test for the landscape character assessment work which has been carried out by SNH and others in Scotland. The most significant part of this will be to explore the extent to which an understanding of landscape can add value to the process of setting long-term goals for the sustainable development of the natural heritage.

3.1   Introduction

Our notion of landscape is the outcome of our visual and experiential relationship with our surroundings. It is an aggregate of the properties of the land, such as landform, land cover and land use, arising through a number of natural and cultural processes, and of our experience of the land expressed as, for example, scale, colour, shape, texture, smell and sound. Thus, landscape as the 'human habitat', can represent a common denominator in a society, despite differing interpretations, values and languages. The study of landscape is thus as much about people as it is about the environment.

Because of this intimate inter-relationship between people and their environment, and the integrated nature of landscape as a subject, its study has much to contribute to the debate on how best to achieve sustainable development.

3.2 Background
3.2.1 Sustainable development
The World Commission on Environment and Development was set up by the General Assembly of the United Nations in 1983 to re-examine the critical environment and development problems of the planet. The findings of the Commission were published in *Our Common Future* (World Commission on Environment and Development, 1987) and based on the over-riding political concept of sustainable development. They recognised that

* meeting human needs involves socio-economic development as well as environmental protection,
* the ability of the biosphere to absorb the effects of human activities imposes natural limits, and
* there is a need to avoid imposing unacceptable risks on future generations.

One of the long-term implications of signing up to, and achieving, sustainable development, is that there will need to be an increased awareness and understanding of the environment, both globally and locally.

3.2.2   Scottish Natural Heritage
The concepts of sustainability and sustainable development have gained ground in both developed and developing countries, linking their needs together within a single unifying concept. In 1992 Scottish Natural Heritage (SNH) was formed through the merger of the Nature Conservancy Council for Scotland and the Countryside Commission for Scotland. The Act of Parliament which established the new organisation required it to undertake its duties in relation to the natural heritage in a manner which is sustainable'. The Act also gave SNH the balancing duty of taking account of social and economic factors (Anon., 1991).

Thus SNH has the challenge of working in an integrated fashion across all areas of its remit (nature conservation, landscape, recreation and access, education and enjoyment), whilst also taking into account the relationship between environmental, social and economic considerations and ensuring the sustainability of the natural heritage.

3.2.3 The Natural Heritage Zones Programme
The Natural Heritage Zones (NHZ) Programme was established in December 1996. Its aims are to integrate all aspects of SNH's natural heritage remit and produce a strategy which takes account of the balance between socio-economic concerns and those of the natural heritage. In particular, the NHZ Programme aims to

* assist in ensuring that national (i.e. Scottish) decisions are made in ways which are sensitive to more local issues;

- assist in ensuring that the present and future needs of the natural heritage of individual local areas are taken into account when national priorities are set;
- assist in the delivery of SNH's statutory duties for the natural heritage;
- target advice and assistance, and the planning of SNH work programmes;
- assist in the implementation of special projects;
- help the public to understand and enjoy the natural heritage; and
- develop with partners (such as Local Authorities and the Enterprise Network) shared long-term visions and action programmes for the natural heritage, founded on the principles of sustainable development.

There are 3 components to this work, namely

- to draw together all information on the natural heritage and the trends of pressures affecting it;
- to articulate the needs of the natural heritage; and
- to build a vision, with partners, for a more sustainable future over a 25 year period.

The information on natural heritage, and the trends and pressures affecting it, is incorporated into three outputs which are designed to influence systematically the building of a more sustainable future. These outputs are the Zonal Prospectuses, the National Prospectuses, and the National Assessments. Landscape, together with other elements of the natural heritage, forms a key component of each.

Building a vision with partners such as Local Authorities, Enterprise Companies, Scottish Environment Protection Agency (SEPA), and NGOs is the phase of the NHZ programme where work cannot be taken forward by SNH alone. This process is multi-faceted with many variables to be taken into account, each with differing priorities. In some zones there may be a pre-existing vehicle for this work such as the Focus on Firths Initiatives or other joint projects, while in others opportunities to develop a shared vision may have to be created.

3.3 Landscape in the NHZ Programme

Nature conservation, through European and UK legislation and its system of international and national designations, carries a number of mandatory responsibilities and a notion of common standards and a common language. Landscape, recreation and non-designated species and habitats, on the other hand, can command varying priorities depending on both local and national policy and its interpretation.

The current progression of the Pan-European Biological and Landscape Diversity Strategy and the European Landscape Convention, through the channels of European debate, may eventually lead to the development of European policies on landscape and to raising the profile of landscape as a subject. However, as landscape does not currently have the same basis of regulation as nature conservation, nor the resulting hierarchy of designations, it has the advantage that it can be accommodated in a flexible manner across the NHZ Programme. Landscape can be a particularly useful means of incorporating a local perspective which is often ignored or perhaps made more remote within a framework requiring an international interpretation. The extent to which landscape will contribute to the NHZ Programme depends partly on the quality of the landscape information and the clarity of the landscape description, but also on the willingness of individuals and organisations to recognise the importance of landscape to society, at local, regional, national and international levels.

3.3.1 The challenge of integration

SNH has a complex remit, as outlined in section 3.2.2. Whilst the overlap between the areas of activity has long been recognised, they have in the past been dealt with in different ways, by different professionals, using different kinds of information. For example, nature conservation is typically seen a reasonably objective natural science, whereas landscape and recreation are social sciences not endowed with the same degrees of objectivity. The differences in information type, information availability, interpretation and ways of thinking between the natural heritage disciplines represent one of the greatest challenges to achieving the integrated outputs which are the goals of the NHZ Programme. A key strength of natural heritage zonation, however, is that it can provide a framework for this integration where the interaction of all factors can be understood to be spatially based.

3.3.2 Deriving the natural heritage zones: the landscape input

SNH's definition of landscape was used to guide the NHZ Programme from its start. The zones shown in Figure 3.1 were derived by combining biogeographic information mainly derived from data on Scotland's geology, climate and natural flora and fauna (Usher and Balharry, 1996) and landscape character information, incorporating cultural, land use and other factors.

These initial zones were drawn up ahead of the completion of SNH's National Landscape Character Assessment Programme. Sub-regional landscape assessments were undertaken by SNH's regional landscape advisors in a systematic but rapid exercise in order to provide the national cover at a coarse scale. Following this initial exercise, and with the availability of more refined landscape information from the National Landscape Character Assessment Programme, boundaries have been re-examined and revised so that the most recent map is derived from the best available data (Figure 3.2).

This national zonation produces a pattern of discrete geographical areas, of generally similar natural heritage character, which face broadly common challenges and opportunities for the future.

3.3.3 Building the working tool

Developing NHZs from a concept to the stage of being lines on a map was not without its difficulties. Taking that idea forward to produce a working tool for the organisation is a major and exciting challenge. In order for the NHZ Programme to deliver its aims, it had to be developed from a basic outline map into an interactive working system.

In order to get a simultaneous top-down (national) and bottom-up (local) approach, it also had to be supported through two different frameworks, national and zonal, so that the national and local pictures could be integrated. The national framework comprises two parts, one providing information and objectives for each key element of the natural heritage (landscape, biodiversity, etc.) and one providing integrated information and forward looking objectives related to series of natural heritage settings (mountain and moorland, farmland, etc.). The zonal framework provides integrated information and localised objectives for each zone. There will be three sets of documents, which were outlined in section 3.2.3.

3.3.4 The National Landscape Assessment
The National Landscape Assessment draws together the following systematic information on landscape, which includes

- the national landscape character data set (see Hughes and Buchan, this volume) at a scale of 1:50,000;
- the individual landscape character assessment reports at 1:50,000 (Hughes and Buchan, this volume) from which the national data set has been extracted and synthesised;
- the National Scenic Area (NSA) designation maps and descriptions;
- the Inventory of Gardens and Designed Landscape. (Land Use Consultants, 1987); and
- landscape designations. in Local Authority Development Plans.

The analysis will be based on computerised data sets and digitised maps, with manual interpretation as necessary, and will be presented as a series of overlay maps, tables and text. Much of the information will be accessed directly from the national landscape character data set and the analysis undertaken is likely to incorporate the following nine components,

- the range of landscape character types nationally;
- the range of landscape character types in each zone;
- an identification of combinations of landscape character types typical of different zones;
- an assessment of the rarity of landscape character types nationally;
- an assessment of the proportion of each zone within NSAs;
- identification of the key characteristics of landscape character types found within NSAs by zone and nationally;
- the number and location of Inventory of Garden and Designed Landscapes sites by zone;
- the number and location of local authority designations by zone; and
- landscape pressures and trends by zone and nationally.

National objectives covering SNH's approach to landscape management will set the strategic framework within which zonal landscape objectives will be formulated, providing a consistent context, without stifling local identity. National objectives for landscape are unlikely to take the form of spatial planning strategies for individual land uses, but will be in the form of strategic principles to be applied or adapted to suit local circumstances.

The production of the National Landscape Assessment relies heavily on the computerised national landscape character data set which constitutes a very powerful tool, although subject to some limitations such as the differential inclusion of coastal areas. However, some information, e.g. Areas of Great Landscape Value (AGLVs) and other regional landscape designations are currently not readily available in a computer accessible form.

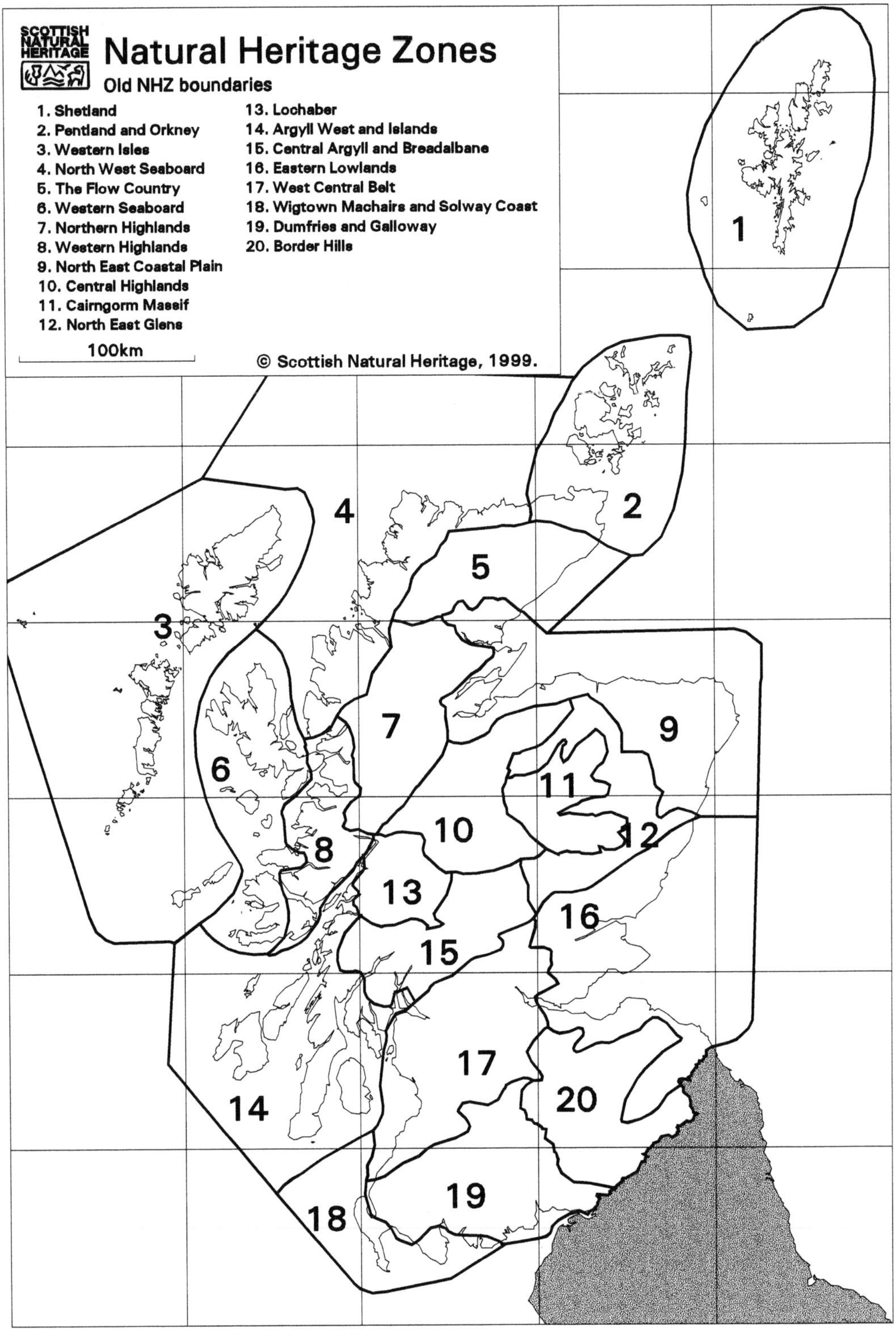
SCOTTISH NATURAL HERITAGE
Natural Heritage Zones
Old NHZ boundaries
1. Shetland
2. Pentland and Orkney
3. Western Isles
4. North West Seaboard
5. The Flow Country
6. Western Seaboard
7. Northern Highlands
8. Western Highlands
9. North East Coastal Plain
10. Central Highlands
11. Cairngorm Massif
12. North East Glens
13. Lochaber
14. Argyll West and Islands
15. Central Argyll and Breadalbane
16. Eastern Lowlands
17. West Central Belt
18. Wigtown Machairs and Solway Coast
19. Dumfries and Galloway
20. Border Hills
100km
© Scottish Natural Heritage, 1999.
1
2
3
4
5
6
7
8
9
10
11
12
13
14
15
16
17
18
19
20

Figure 3.1. Initial zone boundaries drawn in 1996 using landscape character assessment at sub-regional level (1996).

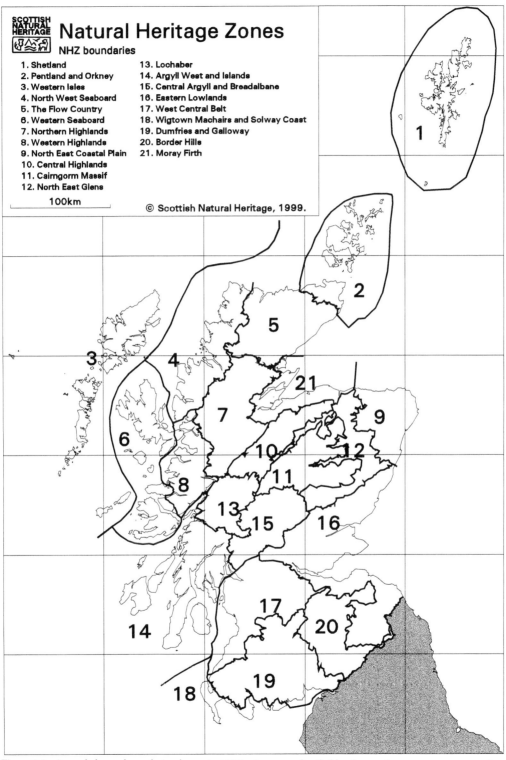

Natural Heritage Zones
NHZ boundaries

1. Shetland
2. Pentland and Orkney
3. Western Isles
4. North West Seaboard
5. The Flow Country
6. Western Seaboard
7. Northern Highlands
8. Western Highlands
9. North East Coastal Plain
10. Central Highlands
11. Cairngorm Massif
12. North East Glens
13. Lochaber
14. Argyll West and Islands
15. Central Argyll and Breadalbane
16. Eastern Lowlands
17. West Central Belt
18. Wigtown Machairs and Solway Coast
19. Dumfries and Galloway
20. Border Hills
21. Moray Firth

100km

© Scottish Natural Heritage, 1999.

Figure 3.2. Amended zone boundaries drawn in 1998 using more detailed landscape character assessment at the 1:50,000 scale.

One of the challenges of the National Landscape Assessment lies in using the detailed information that is provided in the national landscape character data set to produce summary information which is relevant at both a national and a local scale. Another is the identification of strategic principles for landscape management, which although informed by current best practice, are sufficiently robust to influence our long-term vision for the natural heritage. The strength of these strategic principles will be in their clarity, pragmatism, flexibility and capacity for integration with other natural heritage objectives.

3.3.5 *The National Prospectuses*

The National Prospectuses will provide an integrated understanding of the natural heritage and the pressures acting on it at a national level (across Scotland as a whole).

As a framework for securing that integration, six natural heritage settings have been identified (listed in Figure 3.3). These comprise a mosaic that divides Scotland into areas which have predominant characteristics of mountain and moorland, peatland, forest and woodland, urban, coast (including shore and estuaries), and farmland (upland and lowland). These groupings are not (entirely) mutually exclusive. Each setting will contain information on the landscape and landscape characteristics typically found within it.

NATIONAL ASSESSMENTS

- Earth Sciences
- Recreation Access
- Biodiversity - species
- Freshwater
- Environmental data and information
- Landscape
- Biodiversity – habitats

ZONAL PROSPECTUSES

- for each of the 21 zones

NATIONAL PROSPECTUSES

- Mountain and moorland
- Peatland
- Forest and woodland
- Urban
- Coast and shore including estuary
- Farmland

Figure 3.3. Natural Heritage Zone Programme outputs and how they relate to each other.

The effects that trends and changes are having, or will have, on these identified landscape qualities and characteristics will also be described and, if relevant, will be used in the development and analysis of policy for each setting.

Whilst there are no perceived gaps in landscape information, there is considerable challenge in ensuring that suitable distillations of landscape qualities are incorporated into the analysis of each setting. Whilst manipulating the national landscape character data set will provide some of this information, much reliance will be placed on the knowledge and ability of SNH's landscape staff to interpret and integrate the landscape dimension.

3.3.6 The Zonal Prospectuses

The zonal prospectus provides a more localised and detailed analysis together with strategies for the future. Each zonal prospectus is initiated by a workshop where SNH staff working in the zone discuss the zone and its issues. This workshop and subsequent consultation with local partners ensures that a local perspective is brought to the development of policy and strategies. There are five sections to each zonal prospectus.

- Section 1 incorporates a broad understanding of the landscape and its relationship to the socio-economic factors of the zone. It also examines the relationship between the zone and its neighbours and the role that landscape plays in this. The kind of landscape information required for this section is best sourced from individuals with a good working knowledge of the area in question, and with reference to the national landscape character data set for broad landscape changes.

- Section 2 describes the natural heritage of the zone including its landscape component. The landscape types which occur within the zone are described using several levels of information. first, there are the key landscape characteristics derived from the individual landscape character assessment reports and analysed within the National Landscape Assessment, together with local landscape knowledge (especially of the experiential qualities of landscapes, and locations perceived to have a strong 'sense of place' or local amenity value). Second, there is information on nationally and locally designated landscapes, and, third, there may be one off landscape and landscape related studies. It is important to ensure that the more 'objective' visual description presented in the Landscape Character Assessment reports is complemented by a description of the way in which people experience those landscapes and what they mean to them. Empirical information in this area is often very limited, though use is made of local landscape studies where they exist. This information will be presented in a way which clearly indicates and explains the importance of the landscape in the zone and integrates it within an understanding of the wider natural heritage.

- Section 3 includes an analysis of the pressures and trends, forces for change (both positive and negative) in the landscape which are acting on the zone, using information from the National Landscape Assessment and the socio-economic profiles which are undertaken for each zone. This section hinges on identifying the effects on those aspects of the landscape in the zone which have been identified as being important, the cause of those effects, and the inter-relationships with other elements of the natural heritage.

- Section 4 identifies which legislative and ministerial targets (e.g. European Union Directives and UK Biodiversity Action Plan) are to be met within each zone. Whilst at the moment the UK has no such landscape obligations, the last decade has seen an increasing interest in Europe in adopting a more standardised approach to landscapes and it is possible that ministerial landscape targets may eventually need to be built into the Zonal Prospectuses. This section also includes SNH's own targets for the natural heritage and the changes which it considers necessary to achieve sustainable use of the natural heritage. These targets will be established with reference to SNH's wider objectives and the national objectives arising from the National Landscape Assessment. The challenge here is to ensure that targets for the management and change of the natural heritage are integrated across all areas of SNH's remit rather than being isolated landscape targets.

- Section 5 analyses opportunities for the future and provides a vision of what the zone could look like in the future. It will draw from all of the information in the previous sections and is the place for presenting proactive and integrated suggestions for the future management of the natural heritage.

The 21 Zonal Prospectuses are scheduled to be completed by September 2000. Both before and after this date, SNH will consult other agencies and organisations so that the implications and opportunities associated with the NHZ programme, for all those concerned with the natural heritage and its management, are understood and openly debated.

Along with the other dimensions of the natural heritage it will be essential that the presentation of landscape arguments throughout is logical, based on sound information and articulated clearly, if this consultation and tasks arising from it are to be effective. An advantage of landscape as a concept is that it can be used as an accessible base, often at its most subjective, to draw people into dialogue on some of the less well understood aspects of the natural heritage.

3.4 Discussion and conclusions

The Natural Heritage Zones Programme is still in its infancy. Since April 1997 it has developed from a mapping concept to an SNH-wide programme of activity. The role of landscape assessment, its approach and products, in guiding the initial developmental works has been influential and remains so into the current post-piloting phase of the programme. It is one of the few ways in which human appreciation of the natural heritage can be systematically drawn into the NHZ Programme.

The NHZ Programme is likely to have far-reaching consequences. The ultimate aim is to raise the profile of the natural heritage on a par with social and economic factors in both rural and urban situations. An appreciation of Scotland's landscape is common across the rural-urban divide and emphasises the importance of using the outputs of landscape assessment as building blocks to the process of integrating natural heritage considerations into social and economic decision making.

A major challenge in the zones programme now lies in the use and interpretation of the outputs of landscape character assessment. Hence, the NHZ Programme will be a rigorous test for the national landscape character assessment work. The NHZ Programme provides a frame within which natural heritage information can be systematically organised in order to contribute to the debate on sustainable development. The most significant challenge which landscape professionals face, and which is highlighted within the NHZ Programme, is to explore and clarify the extent to which an understanding of landscape can add value to the process of setting long-term goals for sustainable development. Whilst detailed knowledge of the environment will always be important, the new challenge for those involved in both natural heritage conservation and landscape matters is to understand the social and economic value of that heritage and to explain it to the public at large.

References

Anonymous (1991). *Natural Heritage (Scotland) Act 1991.* London, HMSO.

Land Use Consultants. (1987). *An Inventory of Gardens and Designed Landscapes.* Perth, Countryside Commission for Scotland, and Edinburgh, Historic Buildings and Monuments Directorate, Scottish Development Department.

Usher, M.B. and Balharry, D. (1996). *Biogeographical Zonation of Scotland.* Perth, Scottish Natural Heritage.

World Commission on Environment and Development. (NECD) (1987). *Our Common Future: the Bruntland Report.* Oxford, Oxford University Press.

4 Defining the Characteristic Landscape Attributes of Wild Land in Scotland

Dominic Habron

Summary

1.   Wild land is perceived to exist in Scotland and yet there is very little of Scotland which could be termed 'pristine' or 'natural' in terms of land untouched by human influence, a widely used definition of an ecologically 'wild' place.
2.   There is a range of perceived wildness within Scottish upland areas and the perceptions of mountaineers have remained stable over the last 25 years.
3.   The use of a photographic questionnaire helped to identify the characteristics of landscapes which give rise to their perceived wildness. These characteristics can be quantified with the aid of a Geographical Information System in terms of land cover, the presence of human artefacts and the underlying topography.
4.   A predictive model of the wildness of a location based upon the surrounding landscape attributes was developed and applied to areas of the Cairngorms and Wester Ross.

4.1 Introduction

The terminology of wild land in the Scottish context is important. There is no wilderness in Scotland, if the word is understood to describe an area of land, untouched by humankind, in a pristine natural state. However, this is a somewhat simplistic interpretation of the word 'wilderness' although it will suffice for the purposes of the line of argument in this chapter. To the people who live in the Highlands of Scotland the landscape represents an emptied land (Aitken *et al.*, 1995), but one that is perceived more often than not by the urban dweller as an empty wilderness. The degree of human influence on the Scottish landscape has left little untouched, and yet in many places there still exists a sense of wildness. It is these areas that are referred to as 'wild land', an idea which owes its origins to the concept of wilderness.

Fenton (1996, p. 17) has classified North American wilderness as 'primary wilderness', which is "*an area with the full range of its indigenous flora and fauna, large in area and possessing no people or artefacts*". The next level of classification, which is applicable to some areas of Scotland, refers to 'secondary wilderness', defined as "*an area of semi-natural vegetation where wild animals predominate over domestic stock, medium to large in area and possessing few people or artefacts*" (Fenton, 1996, p. 17).

Any label containing the word 'wilderness' implies 'natural' or 'pristine', and hence is not applicable to Scotland. As an alternative, the term 'wild land' can be used in place of 'secondary wilderness' as suggested by Aitken *et al.* (1995) and this practice is followed here. The word 'wilderness' is used in the case where referenced work explicitly uses this term, as is predominantly the case in work from the USA, Canada, Australia and New Zealand, all containing statutorily protected areas of primary wilderness. When the narrative refers to the Scottish context, the term 'wild land' will be used.

Existing statutory definitions of wilderness mainly rely upon measures of remoteness and naturalness which have been incorporated within models to map wilderness areas (Lesslie *et al.*, 1988). Utilising the conventional measures of remoteness and naturalness would highlight very few, if any, areas of Scotland as being wild, and yet it is still perceived as such. The focus of the research reported here was to find the landscape attributes characteristic of those areas of land perceived as being wild in Scotland and then to develop a GIS based model which would allow the location of wild land to be mapped.

Evidence from the USA (Merriam and Ammons, 1968; Osborne, 1980), Canada (Saremba and Gill, 1991) and New Zealand (Kearsley, 1990) indicates that perceptions of wild land are highly variable, and to take a consensus view of the meaning of wild land is one approach that could be used to identify a definition for a particular culture and/or country. However, despite the differing opinions from person to person as to what constitutes a wild land experience, there appear to be several common factors that underlie the concept, namely the enjoyment of nature, an escape from civilisation, a place for relaxation and solitude (Kliskey *et al.*, 1994).

The predictive Geographical Information system (GIS) model developed assigns a wildness value to a particular location based upon the nature of the landscape visible from that point. The rest of this chapter explains how data on peoples' perception of wild land were gathered and how the GIS model was developed.

4.2 Method

A photographic questionnaire was used to gather data on peoples' perceptions of wild land in Scotland. Photographs were taken of the characteristic landscape attributes of the Cairngorm and Wester Ross study areas as identified in the relevant SNH landscape character assessment reports (Turnbull Jeffrey Partnership, 1996; Ferguson McIlveen, 1999). All photographs were taken using a 35 mm Single Lens Reflex camera with a 50 mm lens, which was mounted and levelled on a tripod at a height as near eye-level as possible and the same film type was used throughout the study. To standardise light conditions as far as possible, photographs were taken between 0900 and 1700 on predominantly cloudless days in the summer of 1996. A total of 48 photographs were included in the questionnaire, four of which can be seen in Plate 5; all are listed in Table 4.1. Each photograph contained one dominant landscape feature identified from the landscape character assessments. Respondents rated each photograph in terms of wildness, naturalness and beauty. In addition, other questions dealt with the distribution of wild land within Scotland and the leisure and socio-economic characteristics of the respondents. The questionnaire was administered as a postal survey and sent to four sample groups, all chosen for their direct contact with rural areas of Scotland, namely

- mountaineers, hill-walkers, climbers, cross-country skiers (123): this group was contacted through a list of affiliated mountaineering clubs provided by the Mountaineering Council of Scotland and represented a recreational interest in wild land;

Table 4.1. The location and direction of the 48 photographs used in the questionnaire. An asterisk (*) against a photograph number implies that it was one of the random selection used in the validation of the regression model.

Photograph questionnaire number	X-coordinate (m)	Y-coordinate (m)	National Grid Sector	Altitude (m)	Magnetic bearing (°) to centre of photograph
1	212769	879970	NH	280	315.0
2	297521	797801	NN	600	327.0
3	303355	793339	NO	430	312.0
4	212203	884528	NH	155	277.0
5*	212588	884422	NH	135	242.0
6	200300	877500	NH	525	84.0
7	203620	878372	NH	215	252.5
8	304219	793063	NO	440	339.5
9*	234900	870500	NH	235	312.0
10	289727	807839	NH	250	140.0
11*	203100	878300	NH	260	201.0
12*	289500	811250	NH	255	174.0
13	299200	805100	NH	835	102.0
14	298400	807200	NH	510	358.0
15	296400	799800	NN	1120	40.0
16*	209441	881700	NH	400	267.0
17	210702	883877	NH	195	71.0
18	299600	798700	NN	1235	330.0
19*	212549	884825	NH	155	259.0
20*	302200	808200	NJ	565	180.0
21	297166	802270	NH	865	289.0
22	206000	871800	NH	560	333.0
23*	302242	808309	NJ	570	322.0
24*	195404	890870	NG	20	324.0
25	300662	811558	NJ	400	218.0
26	291777	810777	NH	235	335.0
27	228323	861645	NH	195	282.0
28	302400	803800	NJ	910	321.0
29	196592	893156	NG	10	34.0
30	208900	879200	NH	125	181.0
31	196481	890680	NG	40	148.0
32	298130	803106	NH	915	74.0
33*	217124	875978	NH	220	222.0
34	301900	793800	NO	450	165.0
35	188925	872370	NG	40	54.0
36	204300	878700	NH	170	250.0
37	304279	793580	NO	440	327.0
38	306003	792048	NO	410	256.5
39	306425	791299	NO	425	338.0
40*	304423	793589	NO	440	358.0
41*	187600	887600	NG	80	340.0
42	301952	803296	NJ	795	112.0
43	298027	809481	NH	365	35.0
44*	295988	803961	NH	655	7.0
45*	300622	811483	NJ	400	323.5
46	210878	884172	NH	160	84.0
47	195225	890798	NG	30	146.0
48	291758	810800	NH	230	89.0

- rural inhabitants (145): people who live in the vicinity of the study areas were randomly selected from the electoral register to represent the views of local people;
- rural outdoor workers (148): this sample group included stalkers, farmers and gamekeepers, and was contacted with the help of a publication entitled *Heading for the Scottish Hills* (Anon., 1996); and
- conservation managers (43): the Scottish and Wildlife Countryside Link supported this study by allowing their member organisations to be contacted.

Numbers in parentheses indicate the number of questionnaires returned by each sample group. Other specific groups such as urban inhabitants and foreign visitors were not included in the study due to budget constraints. However, future work with these groups would add further to our knowledge of how perceptions of wild land vary between different groups of people.

4.3 Questionnaire results

A total of 459 questionnaires was returned, a response rate of 47.9%. A statistical analysis of the wildness, naturalness and beauty ratings using the Friedman two-way analysis of variance revealed that respondents clearly differentiated between the three concepts (Habron, 1998). The greatest differences were between the wildness and beauty ratings, then the beauty and naturalness ratings and finally the wildness and naturalness ratings.

The principal influential factors on perceptions of wild land are place of residence, the degree and type of contact with rural landscapes and familiarity with a particular area. On the whole the rural inhabitants group was more accepting of evidence of human influence in wild land areas than the predominantly urban based mountaineers and conservation managers.

Figure 4.1 shows the percentage of the respondents who had visited an area of Scotland and thought it contained wild land. It demonstrates that within Scotland there is a large range of wildness. The idea for this question was taken from Aitken's (1977) work that had

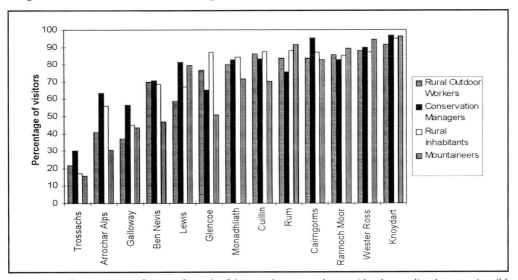

Figure 4.1. The percentage of visitors for each of the sample groups who consider the area listed to contain wild land.

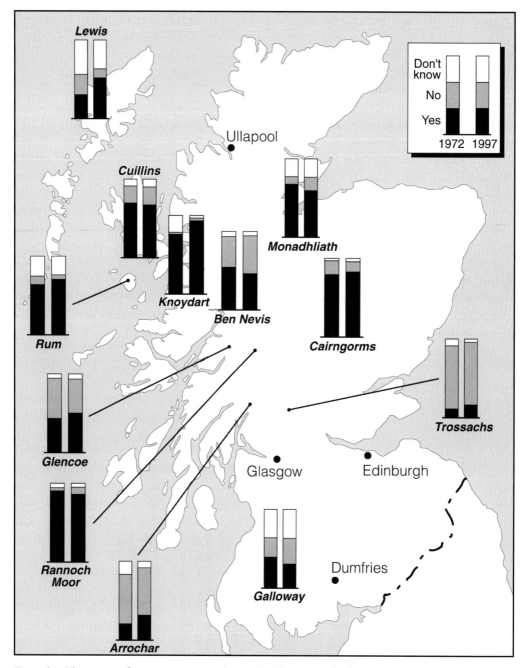

Figure 4.2. The responses from mountaineers to the question 'do you consider there to be wild land in this area?'. The answers are expressed as a percentage of the total number of responses and are displayed for the years 1972 and 1997.

a sample of 748 mountaineers. In order to see how stable perceptions of wild land have been over the last 25 years a comparison of Aitken's (1977) data and that from the mountaineering group in the current study is presented in Figure 4.2. On the whole, perceptions of wild land have changed little over the last 25 years within this group of respondents. For the mountaineering group, wild land areas are characterised by the absence of human artefacts and low numbers of visitors, hence the popular mountain areas of Ben Nevis, the Trossachs and the Arrochar Alps are considered to be less wild than areas such as the Cairngorms, Knoydart and the Cuillins of Skye.

4.4 GIS wildness model

A visibility analysis was conducted to recreate the area visible (the viewshed) in each of the 48 photographs used in the questionnaire (Table 4.1). Three visibility zones were used in the analysis with the following labels: foreground (up to 250 m), mid-ground (250-750 m) and background (beyond 750 m). Baseline topography and land cover data were provided by the Scottish Office and the analysis conducted using ARC/INFO. Landscape attributes including area of land cover, length of footpaths and tracks, number of buildings visible, and area of slope categories were quantified for each viewshed. Multiple linear regression analysis using data from a random sample of 34 of the photographs between the landscape variables and the photograph wildness ratings, allowed a wildness model to be developed which explained 60.5% of the variation in the photograph wildness ratings in terms of the surrounding landscape attributes (P ≤0.05). The remaining unexplained variation in wildness ratings might be due to landscape attributes which were not measured in the study. Independent validation of the model with the remaining 14 photographs (see Table 4.1) gave a correlation of 0.7 (P ≤0.05) between observed and predicted wildness scores.

In the wildness model, shallow angled slopes and hill roads had a negative influence on wildness, whereas steeply angled slopes, higher elevations and the amount of heather and peatland visible had a positive effect on wildness. The area of montane habitat and cliffs in the background of a photograph had a very small negative effect on wildness which could be due to the shallow angled nature of much montane habitat in the Cairngorm area. Unfortunately there are no other statistical models of wildness with which to compare the work from this study.

The model was applied to a 10 km × 10 km section of each study area at a grid resolution of 500 m, giving a total of 400 cells. The output from the model was normalised on a scale of increasing wildness from 1 to 5. Semivariograms (graphical output from a test of the spatial variation of the predicted wildness values) of these results indicated that the rate of change in wildness over a given distance is greater in Wester Ross than in the Cairngorms, and this is attributed to differences in their landscape character, principally the underlying topography.

4.5 Conclusion

The GIS wildness model presented in this paper is a first attempt at the problem of creating a definition of wild land based upon the landscape attributes visible from a particular location. As such it should be viewed as a preliminary model that would require further development and testing before it could be used as a decision support tool for the criterion of 'wildness' in selecting areas for protection, for example as part of a National Park system for Scotland. Future refinements to the model could test the influence of river systems on wildness ratings and could seek the views of a larger range of people. This research has shown that much of the Scottish Highlands, in particular the upland areas, is perceived as being wild despite the fact that very little ecological or pristine (in terms of the lack of human influence) wild land remains.

Acknowledgements

I wish to thank Dr. Ian Simpson and Prof. Donald Davidson for their supervision of this project. In addition, my thanks go to the Mountaineering Council of Scotland and Scottish Wildlife and Countryside Link for their help in sending questionnaires to their members and to the Scottish Office for supplying the digital map data needed for this project.

References

Aitken, R. (1977). *Wilderness Areas in Scotland.* Unpublished Ph.D. thesis, University of Aberdeen.

Aitken, R., Watson, R.D. and Greene, D. (1995). *Wild Land in Scotland: a Review of the Concept.* Unpublished report.

Anon. (1996). *Heading for the Scottish Hills.* Glasgow, Scottish Mountaineering Trust and Edinburgh, Scottish Landowners Federation.

Fenton, J. (1996). Wild land or wilderness - is there a difference? *ECOS*, **17**, 12-18.

Ferguson McIlveen. (1999). *Ross and Cromarty Landscape Character Assessment.* Scottish Natural Heritage Review 119.

Habron, D. (1998). Visual perception of wild land in Scotland. *Landscape and Urban Planning*, **42**, 45-56.

Kearsley, G.W. (1990). Tourism development and users' perceptions of wilderness in southern New Zealand. *Australian Geographer*, **21**, 127-140.

Kliskey, A.D., Hoogsteden, C.C. and Morgan, R.K. (1994). The application of spatial-perceptual wilderness mapping to protected areas management in New Zealand. *Journal of Environmental Planning and Management*, **37**, 431-445.

Lesslie, R.G., Mackay, B.G. and Preece, K.M. (1988). A computer-based method of wilderness evaluation. *Environmental Conservation*, **15**, 225-232.

Merriam, L.C. and Ammons, R.B. (1968). Wilderness users and management in three Montana areas. *Journal of Forestry*, **66**, 390-395.

Osborne, J.R. (1980). *A Scaling Model of Human Impact in Wilderness-like Areas.* Unpublished Ph.D. thesis, Indiana University.

Saremba, J. and Gill, A. (1991). Value conflicts in mountain park settings. *Annals of Tourism Research*, **18**, 455-472.

Turnbull Jeffrey Partnership. (1996). *Cairngorms Landscape Assessment.* Scottish Natural Heritage Review 75.

5   LOCAL DISTINCTIVENESS IN LANDSCAPE CHARACTER

Roland Gustavsson

Summary

1.   Distinctiveness in landscape character is to many an expression with a positive meaning, related both to what happens in the physical landscape and to the way that we perceive the landscape and its changes. Local distinctiveness is focusing special qualities at the local landscape level.

2.   The landscape of tomorrow will be a more planned landscape. Landscape planners are important today, but they will be even more important in the future so that high quality landscape character and local distinctiveness will persist.

3.   Through history the dominant view of landscape character has been to regard landscape as something static, frozen in time. With increased influences of an urban culture, it is very probable that these phenomena will create problems. How can we help planners and land managers to understand dynamics better?

4.   In order to meet the problems related to landscape change there has been considerable activity from conservation bodies in making different types of inventories as a basis for an inventory-led approach to 'planning'. However, identifying the values is just a limited aspect in the planning process.

5.   Search for generality might result in standard methods and simplistic landscapes, where good basic ideas are reduced to clichés. How can we avoid making similar mistakes to those of previous decades? How can GIS help in the planning process?

6.   A reference landscape can help us to achieve a knowledge with necessary width, depth and integration, corresponding to identified landscape character categories, etc. But, what can we gain by using reference landscapes - locally and regionally, including comparisons along a west-east geographical gradient?

7.   Few people question the need for links to historical reference landscapes, but we also need to go beyond studies of the past. What could we gain by the introduction of landscape laboratories as an approach?

8.   'Object-based' concepts are dominant today. It is very different from a 'place-concept'. How can we achieve a 'place concept' and 'local distinctiveness'? Two case studies are discussed.

5.1 Introduction

5.1.1 Local distinctiveness in landscape character

Distinctiveness in landscape character is to many an expression with a positive meaning, related both to what happens in the physical landscape and to the way that we perceive the landscape and its changes (Brinckerhoff Jackson, 1994; Crowe, 1995; Appleton, 1996). Local distinctiveness is focusing the special qualities at the local landscape level, investigated very much from a 'place concept'; qualities which are for landscape character what a backbone is to a man. The qualities belonging to the landscape character are related to natural landscape features, cultural objects from different times, but also very much to the activities, historical events, legends and more recent stories told about a place or a local landscape. People are said to be very sensitive to the kind of signals the landscape character gives; if it signals care, generosity, a long term sustainable use, a linkage to traditional patterns, health, attractiveness, or if it signals carelessness, greediness and exploitation (Ward Thompson, 1998; G. Sorte, pers. comm.).

To improve or preserve local distinctiveness in landscape character will certainly be a challenge in a fast-changing world, with increasing influences from an urban culture, with a 'need' for simplistic standard methods, with decreasing possibilities of keeping a differentiated deep knowledge alive about landscape management concepts, or with difficulties in keeping a bank of articulated reference landscapes, which allow us to understand previous landscape qualities and create possibilities for the future.

5.1.2 Landscape changes, managers and outsiders

The landscape around us is always changing, and it is changing dramatically. What has happened during the last three or four decades has created more environmental problems than any period since the last glaciation. The most striking rural changes in Sweden are the extensification of the small-scale mixed forest-agriculture landscape and the intensification of the agriculturally-dominated lowlands. The existence of whole local 'islands' of agriculture and rich broad-leaved forests in the forest-dominated regions are threatened in the long term, despite new and better environmental support since Sweden has joined the EU. The money helps these to survive but gives few possibilities for future investment. It was still possible to find agriculture in the archipelagos in the 1960s and 1970s, but it is almost absent today, despite its importance for tourism and national character.

People and landscape are more and more isolated from each other. This is a process that has happened step by step. Early settlements were very much related to features in the landscape and from these derived a local distinctiveness based on logical relationships between cultural features, such as buildings and roads, and natural-cultural patterns related to land use, field patterns, etc. Old place names very much reflect these facts. When the villages in Sweden were split in the land reform that took place about 150 years ago, individual farmers were allocated new isolated places in the surrounding landscape. The new locations were more often decided from mathematical-functional criteria, based on local transport, rather than from criteria such as finding a new place with striking landscape features, beauty and good local climate.

Separation is also reflected in how few people directly work in the landscape. Since the Second World War the number of landowners working as full-time farmers or foresters has decreased dramatically. Today the figure is less than 20 percent in Sweden. Furthermore, many houses and plots are sold and separated from the surrounding landscape, with the consequence of less management interest.

There is an increasing need for landscape planners. In the future this is likely to result in a smaller chance of keeping a thorough and local insight about landscape and landscape management among landowners and local people. In parallel, Governments will continue to introduce new policies (e.g. biodiversity, sustainable management, Agenda 21) which will in many cases need planners as initiators and co-ordinators. So, altogether, if we in the future wish a high quality landscape character and local distinctiveness to persist through developments, it also means an increased need for help from outside, including an increased need for landscape planners. If this is so, it will be of crucial importance that these 'helpers' are well trained, and understand the special conditions of each locality, both in landscape terms and socially. There will be an obvious risk of uniformity, based on some objective scientific criteria on a national level, driven into a simplified planning process, processing all land as equally as possible, forgetting about the complexity bound to the local identity of a place and the local history.

In order to meet the problems there has been considerable activity from the conservation interests making different inventories as a basis for an inventory-led approach to planning. Within these inventories, the 'landscape values' are said to be found. However, will they stay as 'values'? How they develop will very much depend on the direction given to management objectives. Identifying the values is therefore just a very limited aspect in the planning process. Consequently, the inventory now completed by Scottish Natural Heritage on the landscape character in Scotland is of great value (see Hughes and Buchan, this volume). From a landscape planning perspective something similar would also be very useful in Sweden. Many of us are now looking forward to the next steps in Scotland, focusing on how to implement and integrate landscape character into planning and management.

5.2 A time perspective: static or dynamic

The dominant view of landscape character in the past 200-300 years has been a view of landscape as a picture, frozen in time. With an increasing urban culture it is very probable that the view of the countryside landscape as being static will increase. This can create problems. In dealing with landscape, landscape character, planning or conservation issues, it is necessary to examine past, present, and future situations, raising questions about how we choose a time reference. How can we improve our ability to have a dynamic rather than a static landscape? How can we help planners and land managers to understand dynamics better?

Early in this century the French philosopher Bergson greatly influenced the search for knowledge. Today he is probably unknown, but many are rediscovering his work. The heart of Bergson's criticism concerned the 'traditional philosophy' which he claimed forgot the essence of the matter, the qualitative matter, focusing too much on quantitative aspects. Bergson thereby stressed the importance of a dynamic perspective. Some quotations from Bergson (1928) are 'The absent-mindedness of the qualitative differences is particularly a matter of distortion of the understanding of time'; 'Reality is a question of movement.

Reality is the uninterrupted change of forms. The form is just an on-the-spot account, taken from a transition'; and 'There is a clear relationship between motion and degree of consciousness'.

Studies of landscape change, especially relating to the question of quality, are important today, particularly when seen in a landscape planning and educational perspective. Change is mentally problematic to deal with for all human beings. It is an important challenge, to take Bergson´s view seriously, using research and other studies to increase our awareness of qualitative dynamic aspects. Eight aspects can be explored.

First, many studies have compared different historical maps for the same local area (Ihse, 1995; Skånes, 1996). With help from computers, it is now possible to go further, so that each planner has the possibility of seeing earlier phases as special time-layers superimposed over the picture of the present situation. However, having information about the past does not guarantee that the dynamic perspective will be applied. Furthermore, there is a need for studies encompassing planning, implementation and management process from dynamic perspectives.

Second, historical sources concerning landscape are generally good in Sweden, especially around the middle of nineteenth century. Unfortunately, the same sources are not available for the eighteenth century, when the numbers of meadows, wooded meadows and probably coppiced woodlands were much greater. That was also the time when the pastoral grazing system was abandoned over huge land areas. There is also little information about the landscape earlier this century, which in many ways is the root of the landscape of today and therefore of special interest for planners, and when considering the interpretation of local identity. Consequently, several researchers in Sweden have wanted to put the twentieth century in focus, trying to improve the knowledge of the 20th century by answering questions about cultural identity, characteristic landscapes and traditional management methods when differentiating time into different periods within the 20th century (Gustavsson, 1986, 1988, 1991; Gustavsson and Fransson, 1991; Sarlöv-Herlin, 1991; Bucht, 1997).

Third, our generation is just watching the very beginning of the consequences of modern landscapes and management (Gustavsson and Ingelög, 1994; Fry and Sarlöv-Herlin, 1997). Individual species can still look vital but their long term ability to live or reproduce themselves might have gone. There is a time delay of the effects of changed management methods appearing; we should be aware of this. This is shown in several landscape ecology studies, a phenomenon sometimes named 'the landscape memory'. The time delay could also be discussed in terms of preferences and attitudes, including those of planners (Turner, 1996).

Fourth, the argument has been about the importance of letting historical knowledge run right up to the present day. To go further, and to reach the challenge of Bergson, we also need to expand the time view within our own time, to be able to understand that the landscape changes throughout the whole of time. Husserl (1995) used terms for the immediate experience of what has just happened (retention) and for the immediate experience of what is just going to happen (protention). By this extension of time we get a chance of a time-flow experience, expanding the static present to include before and after states.

Fifth, landscape character presents basic features for the future. Historical analysis can be a part, showing links from the present to the past. Examination of current and past landscape character can thus give important guidelines for future landscape character. However, it is a question of more than just history. The dynamic forces acting on the landscape make everything other than change unrealistic. It is probably only in a few exceptional cases that 'a frozen landscape character' will survive into the future.

Sixth, on a more fundamental level it does not mean that the past will, or should, lead us forward. The forces or motives driving us forward can perhaps be found in the past, in our roots, but should probably even more be searched for in things which lay ahead of us. This is stressed by the philosopher Popper in a reflective lecture given at the age of eighty-four. He said that 'The future is open, remains to be resolved. Only the past is set. It has been implemented and is thereby passed. The present on the other hand is an ongoing process.' and 'In the view of the old world the past mechanically pushed us forward. But it is not the force or the kicks from behind which make us move; it is rather the temptation of future that attracts us or drives us forward' (Popper, 1996). The search for future concepts and future landscape values should therefore not be limited to involving conservation and reconstruction issues only. Landscape character should inspire us in a wider context.

Seventh, considering sustainability and time perspectives, we need to be careful if we believe that the increased direction towards both environmental and sustainability issues is satisfactory. If the 1960s and 1970s are the reference time that we use, it might be true, but then we are comparing today with probably the most significantly exploitative period in human history. Instead, we should perhaps use the 1940s or the 1950s as a reference time.

Finally, if we want the landscape to be alive in our society, we need a mental landscape, that it is possible to formulate in thoughts and words. If we want to be certain about possible future developments and management strategies we must be able to formulate concepts, conveying the mental landscape in a correspondingly precise and nuanced way. Consequently, local distinctiveness should also include places related to both stories and personalities behind the landscape. We also should combine landscape studies with a semantic approach to research, focusing on local names of places as well as the deeper meanings of landscape terms. To illustrate the latter there is an ongoing research project in Sweden concerning the term 'lund', one of the central landscape terms in the Nordic countries, but today relatively unknown compared to terms like 'meadow', 'wooded meadow' or 'garden'. The term 'lund' is concerned partly with small, tree-rich places situated close to old settlements on former inbye fields; today these are botanically rich in species, and with parallel richness in cultural background. As a landscape term it has relationships to English terms like 'grove' and 'ancient woodland'. It is a living term, changing in a fascinating way in relation to time and context. As a result, a better understanding of the landscape term 'lund', and the ability to interpret local landscapes, will increase. In parallel we can be influenced when trying to identify new development concepts (Gustavsson and Wittrock, 1997). As Collingwood (1939) said 'If we are to study how people think we must look and see in what terms they think'.

5.3 Beyond the past: towards the future

History is necessary for us to be able to understand landscape processes, or when interpreting landscape character qualities, or in understanding the sense of place. We need history but should not be bound by it. Using an expression of the Danish philosopher Kierkegaard: no new generation should be forced to go backwards into the future. Consequently, what ways are there of using existing landscapes and going beyond the past? Three approaches are discussed here: the landscape laboratory concept, the case study concept, and studying countries in other parts of the world. All three are, but in different ways, raising the question of how to relate to the past without aiming to copy or freeze it; how to go beyond the past; and how to pass the limit of purely empirical studies about what currently exists. To do this there is a need to stress 'live' research, investigating landscape character, focusing development and reconstruction concepts, as well as a need to decide how to go from a descriptive, inventory-led and analytic-led research to a more experimental, interactive, solution-directed research. Two of the three alternatives are here illustrated by ongoing research programmes at the Swedish University of Agricultural Sciences.

5.3.1 The Alnarp Landscape Laboratory

In 1990, the Faculty of Agriculture at the Swedish University of Agricultural Sciences, Alnarp, decided to set up a landscape laboratory (Plate 6). With its focus on the construction of new elements and landscape character, it is unique in Sweden. Landscape elements, which are part of every day design and landscape planning, such as small woodlands, groves, woodland belts, hedges, avenues, meadows, streams and small water areas are highlighted for research and demonstration in such a way that a whole series of main- and sub-types can be studied intensively side by side, as well as being part of specific landscape patterns.

There are striking similarities in the physionomy and ecological conditions for these 'key-stone elements' when comparing agricultural areas with open spaces between residential or industrial areas regarding size, interior structure, extended shape and isolation problems. These similarities are used when trying to combine aesthetic qualities with good conditions for a rich flora and fauna and a favourable local climate. The landscape laboratory provides opportunities to continue the successful research carried out at Alnarp during the 1970s and 1980s on vegetation structure, biotechnical methods for creating new habitats, environmental psychology and landscape analysis methods, as well as agricultural and forestry management. Five particular aspects of the research are mentioned.

First, there has bean a focus on testing new development concepts. The landscape laboratory gives researchers new possibilities for comparing alternatives, based on strictly identified concepts which are not subject to the usual land-use compromises. In this way it is meant to be a supplement to other forms of research, based on empirical studies in existing, traditional landscapes, supporting researchers to go on beyond past. However, it is important to underline that when identifying new concepts, the aim is usually to have links to other landscapes, history and traditions by using 'key reference landscapes' and to characterize the transformation process in an explicit way. How do we go from an old, mature landscape character to a planting scheme for a new one, making the transition to the basic qualities as soon as possible, and how can we show the ideas in an explicit way (Gustavsson and Rizell, 1998)?

Second, there can be a greater insight and a deeper knowledge base. The landscape laboratory gives an opportunity to deepen and broaden knowledge, bringing different aspects together concerning the same geographical landscape. This can lead to a more integrated and a more fully understood knowledge, as well as a knowledge where the context and spatial patterns are stressed.

Third, there are opportunities for full-scale experiments. These experiments, laid out to work over a time-scale which is necessarily long, will provide an opportunity for understanding processes and interactions in a dynamic way. By constructing new prototypes these can be tested, new hypotheses can be formulated, and adjustments can be made before setting up larger series or applying the results on a large scale. This should lead to an ability to see alternatives and new possibilities in landscape design and management planning, and should therefore be an important tool when aiming to create landscapes with a distinct character, a strong cultural identity or habitats of high quality.

Fourth, it is possible to concentrate on one specific, favourably-situated landscape. The opportunity to walk from one alternative to another, and to be able to make direct comparisons, is important. The close connection to the University campus of Alnarp gives students, researchers and teachers excellent opportunities for repeated contact over the seasons and over the years. The main studies on the 300 ha estate of Alnarp will, however, be supplemented by other locations in the vicinity. Today such 'satellite areas' are at Snogeholm, focusing on the design of combined environmental-production directed plantations, and at Toftanäs in Malmö, focusing on the creation of new wetlands and small water bodies.

Finally, the landscape perspective can be explored. The project considers landscape elements and habitats which make a central contribution when constructing or reconstructing agricultural and urban landscapes. Research aims to enrich these landscapes for their combined aesthetic, flora, fauna and local climate qualities. The focus is on the spatial configuration affecting human preferences, links to cultural and natural heritage, and the survival and distribution of species. Different patterns for different parts of the Alnarp Estate are introduced, thereby making it possible to study connectivity and continuity aspects as well as landscape character and distinctiveness at a local landscape level.

5.3.2 The Skrylle Recreational Forest project

The second approach is to use case studies. The example used here illustrates a mixed 'problematic-successful' case study (Flyvbjerg, 1991). The project is concerned with forest recreation and forestry planning in an area in the south-west part of Scania, Sweden, focusing on questions about landscape management planning, restoration and landscape character concepts, at a series of different scales. The area is statistically the second most visited recreation area in Sweden, with about one million recreational visitors each year. Despite the large number of visitors, most investment has been made for intensive timber production and 'natural values'. From the project experience so far it would be useful to address the following six problems.

First, there is the problem of how to use geographical information systems (GIS). The main aim in the Skrylle Recreational Forest project is to improve the understanding of how goals and strategies go into action by focusing on the management plan and the implementation process. A series of maps has been drawn through the years in order to describe different kinds of landscape values and goals. Most recently a 'nature conservation-directed forestry plan' was taken through from the early 1990s to a suggested plan for 2005. Some of the strategies are formulated in words in the management plan, but there are also unwritten strategies to be found, which might be even more important for eventual realization. When found and described, the challenge is to reveal how they could be made more transparent in a participatory planning process, and hence help to integrate the spatially-related issues in a graphic manner. In Sweden, GIS has mostly been used in the inventory phase. How can GIS be developed to be a better instrument in the implementation process? A lot of information can be collected, but how can the mistakes of the 1960s be avoided? At that time the interest was in developing a landscape-based planning system, but it failed, partly because an 'ideal plan' was far too focused, and partly because there was too much information presented in an unstructured way.

Second, there are the problems associated with flexibility in time and participatory planning. One of the main problems of today, and in the future, is the primacy of the timber industry. It is not a surprise that the increased need for recreational areas was not anticipated in the 1960s. Other anticipations of the 1960s are now required to have greater flexibility in the landscape planning process. By participatory planning it might be possible to find out about the popular or problematic places of today. More problematic, but just as important, is the identification of networks, and of relationships between places and networks.

Third, do you approach landscape from the top downwards? Many people argue that a landscape should be analysed and planned from the top downwards, giving priority to solitary full-crowned trees, a small hill or a stream. Others argue for a bottom upwards approach. Maybe a combination of the two approaches is the way forward in the future. Starting from the bottom upwards, the small places creating a strong identity can be taken into consideration; these are also part of a landscape character at a larger scale.

Fourth, there is the problem of the place concept. The Skrylle project is a case study in which a place concept is tested as one approach. A place concept is one that regards a whole local landscape, with its entrances, characteristic landmarks, as well as its physical, visual and mental passages, dominant and unique landscape elements, and its relationship to its surroundings. It therefore differs from the most popular landscape concepts of today among ecologists and foresters, as well as among many representatives of cultural heritage organizations, in which the individual objects, habitats or forest stands are isolated phenomena. Preference studies have by tradition detached the habitat type or woodland character from its context, as occurred in the earlier studies of the Skrylle area. But the landscape context seems very important, a fact that the following example will illustrate. Early in the restoration process ecologists wanted to eliminate all spruce in the Skrylle area, but interviews found that there was an interesting spatially-related preference based on home and a sense of distance. The spruce represented the distant and far away feelings, indirectly making open grazed areas and broad-leaved dominated places to be regarded as special places, the home places (C. Fredriksson, pers. comm.).

Fifth, what is the importance of entrances? The assumption was that the first impression of an area was the most important. From that perspective all important entrances to the landscape were categorized and the potential major entrances of tomorrow were identified.

This analysis gave useful results; if the assumption is true, most entrances should be improved.

Finally, there is the problem of story telling and the hidden mental landscape. Researchers such as Söndergaard Jensen and Koch (1997) have shown how information affects preferences of landscape. A few years previously, a restoration plan for another recreation forest, 'Furulunds Fure', was developed. Part of the work investigated 'a hidden story' belonging to the forest in the early part of this century. By finding and promoting that story the forest was lifted from a forgotten place to the forefront of people's perceptions, highly appreciated in the society by both local politicians and several interest groups. This created an unusually high level of support for the restoration plan (Gustavsson and Fransson, 1991). For Skrylle no corresponding story was found, but the aim was to get to the point of suggesting future directions in a management plan. This is normal, but should it not be questioned? We are inclined to follow the traditions of the natural sciences, always trying to think objectively, always keeping to hard facts, and reducing language to mathematics. There is a shortage of scientists as story tellers. Indirectly it also decreases the possibility of giving a meaning to the knowledge in an integrated knowledge process. C. Foster (pers. comm.) argued about aesthetics and forests, saying that the most important task that the researcher has is to be a story teller. To tell true stories plays a significant role in making the mental landscape grow, for both experts and public alike.

5.3.3 Comparisons with North American landscapes

A slightly more daring alternative is to make comparisons with existing or historically documented landscapes in other parts of the world. The northern parts of the USA and the south east parts of Canada are used as examples, including areas like Yellowstone, Black Hill, Minnesota, northern parts of Ohio and Pennsylvania in the USA, as well as Nova Scotia in Canada. These are areas which, when compared to southern Sweden or Scotland, have a similar climate, a similar glacial history, and also a corresponding ecological situation straddling the zones of spruce and broad-leaved forests. There are of course many differences, especially as it would be difficult to find the same spectacular landscapes in Sweden. However, what is striking in such a comparison is the fact that the physionomy and dynamics of the landscape can sometimes be very similar, despite a different set of key-stone species. The following are four comparisons.

- In Sweden, researchers have recently found that oak dominance is due more to former land use and settlement patterns than to macro-climatic factors. In the USA a similar relationship with the native American Indian settlements and management has been shown.
- This leads us to the question of what Northern European landscapes would look like if the agricultural culture had not come so early, destroying earlier human cultural patterns similar to the North American Indian cultures. What would Sweden have looked like if we had maintained the pastoral grazing system with shepherds beyond the late seventeenth century? Looking for future approaches to landscape management, these concepts would be interesting because they are certainly part of history, even if written documents are missing. Related large scale grazed landscape patterns, fluidly shifting between more open and more closed in character with trees and shrubs, would today be

considered to have many basic aesthetic qualities. At the same time they also provide attractive habitats for many threatened species, which have succeeded in surviving in the agrarian landscapes.

- The auroch cannot be part of the future, but maybe the European bison and the European wild horse could be. From a landscape point of view, it is interesting to figure out which landscape patterns, physionomy, character and dynamic processes might occur. Some experience can today be found in Eriksberg, situated in the south Swedish archipelago. The area, covering 1,000 ha, has been grazed by large herbivores, including the European bison, since 1930. This provides an example, but it also indicates the need for more thorough research.

- Finally, in America it is surprising that there appears to be so little interest in the eastern cultural landscape, including the tall grass prairie landscape.

5.4 The use of reference landscapes

The importance of a linkage to reference landscapes has been stressed in most of the earlier discussions. Reference landscapes, as a fully developed concept, could be a very effective way to give the necessary integrated and deep knowledge about processes, elements and patterns related to landscape character. Identified types at regional as well as local level could be more fully understood in their context, related to time, space, use and cultural meaning. It is improbable that we shall succeed in identifying interesting concepts for future landscapes without any reference to existing ones. It is rather like someone who wants to be a good writer, but has no access to a good library. This means the rediscovery of an earlier tradition, as for example Lindquist (1938) and Selander (1955) in Sweden, Rackham (1975, 1976) and Peterken (1981) in Britain, and more recently Whitney (1994) in the USA. It also means that there is a need for an improved philosophical and theoretical basis, formulated to make it a respected area of research. If we want to pick some facts here and there, we may advance one or two steps forward, but not far from a landscape knowledge point of view. Hence, a more systematic search for good reference landscapes should be an important aim. The question then arises of how we search for these reference landscapes, and maybe also how to search for new ways of conserving or recreating them as well as new ways to understand them.

Natural science stresses a search for generality, but does generality really help us to find interesting solutions for the future? If we search for generality we can come to conclusions about the state of the art or about how common problems are. This is important but does not necessarily reveal anything about what might be good development concepts for the future. When searching for novel development concepts should the study rather be directed towards several unique areas? If we do so we will be using an approach more common to human science.

Natural science helps us to get good answers to individual questions. An integration of traditions from natural science and human science might give new possibilities of enclosing facts within essential contexts. We need to apply a more holistic landscape view and pick out a series of reference landscapes, in time and in space, at international as well as national, regional and local levels. The reference landscapes should be selected so that a few are prioritized as key reference landscapes, while the majority should be regarded as providing the necessary width (see Plates 7, 8 and 9).

Reference landscapes should help us to get a more fully integrated knowledge of landscapes both as places and as specific patterns belonging to a local distinctiveness. If we stay too close to home, our concepts will be too narrow and we will have difficulty seeing our own unique qualities, i.e. the local distinctiveness. Searching for concepts for a fuller understanding of landscapes may sometimes mean that it will be more productive and informative to study landscapes other than those that are most familiar. Sometimes you should go to another country, for example to study a Swedish landscape rather than an additional Scottish one! The following two sections provide examples.

5.4.1 A west-east reference line

Sweden covers many climate zones from north to south. In the southern part of Sweden, there is a landscape where the deciduous zone meets the coniferous zone, creating an interesting 'tension zone' in a European context. Here ecosystems, species and cultures try to expand or struggle to exist. There is also a west-east climate line found here, with a more humid and maritime climate in the west and a more continental climate in the east, similar to a gradient that can be found across Scotland.

The differences in climate have rarely been discussed but are undoubtedly significant for the Swedish landscape characteristics. In the western part cattle may be grazing outside all the year round, meaning a greater grazing pressure and a more open landscape with less continuous tree cover. In the eastern part the cattle have to be stabled during the winter months, resulting in less grazing pressure and land with abundant trees and shrubs. This led to a more diverse landscape with a higher degree of tree cover and greater connectivity.

Working in the southern part of Sweden it is interesting to extend a west-east geographical line, searching for reference landscapes all the way from Scotland to the Baltic countries. This allows us to see alternatives and to get a deeper understanding of the 'home landscape', based on investigations in Estonia and Scotland. Interesting themes for individual studies along such an expanded west-east reference line might be the pastoral landscape (Gustavsson, 1995). You also become aware of the rich natural and cultural countryside heritage, coupled with a strong local distinctiveness, in landscape character that still exists in many local landscape areas. These landscape characters are very important as references for future concepts. It also becomes easier to understand earlier phases in the south Swedish landscape. An example would be the forested landscape of today in the south-east part of Sweden; earlier this century it was described as being dominated by juniper, a fact which is difficult to understand today. In Estonia similar phases are common and thereby possible to study. The agrarian heritage in Estonia is currently threatened; if you want to study an agricultural sector that has been totally regulated, and its recent change when there has been no regulation, environmental support, or other help, this is essentially the situation in Estonia. This evokes a discussion about the joint responsibility of conservation between countries.

Culturally, eastern Sweden is said to be influenced very much by the historically rich Baltic cultures. The western side of Sweden has correspondingly shared cultures with societies in the west of Europe, which today is apparent in architecture, types of fence, etc. Later, the countries have gone in different ways, but that makes it perhaps even more interesting to compare the different countries.

Modern lowland agricultural landscapes, like those found in the surroundings of Edinburgh or in southern Sweden, are being pushed towards even greater intensification.

This can cause problems with greater pressure on the road networks, as well as on isolated elements such as small woodlands, hedgerows and diminishing public accessibility. Loss of cultural patterns and local distinctiveness is a frequent result.

To compare the low mountainous areas of southern Sweden and Scotland highlights differences, but also helps us to consider similarities related to tree continuity, rarity of trees, and the diversity of tree and woodland systems. In both Sweden and Scotland there are similar interests in the restructuring of the very simplistic spruce plantations which are not in harmony with the landscape. Sweden now has much greater tree continuity and woodland system continuity. Frequently in the Swedish landscape you will find examples where the natural features and semi-natural systems have been allowed to return, or are being reconstructed in a forested landscape which is still used for timber production.

Sweden has come to a point where the agrarian heritage is part of the forestry culture. The case is often made that the last farmers have left and the only possibility of taking care of agrarian remnants are people and organisations belonging to a forestry culture. In many other European countries the transformation to a forested landscape is currently happening, and there is concern about how historical features belonging to former agrarian cultures can be handled after the transformation. The question is whether this can be achieved without losing local distinctiveness.

5.4.2 Long term research areas as key reference landscapes

The most important reference landscapes should be the focus of long term research, covering many aspects including the changes over time. Long term research areas, such as the Alnarp Landscape Laboratory and the Skrylle Recreation Forest should be good examples of what could be considered as reference areas of considerable importance, i.e. key reference areas. Tjärö, an island with major tourist and conservation values, could be a third one. Every year for at least the last ten years it has been used for educational programmes to train landscape managers (Gustavsson, 1997; see Plate 10).

The grazed landscape with its traditional pastoral characteristics could be a fourth landscape type. In Sweden it is considered to be one of the most important landscape types. For flora and fauna values, about 80 percent of the threatened species are said to have their major living space within these types of landscapes (Gustavsson and Ingelög, 1994). The pastoral landscape is also said to have many highly appreciated, classic, aesthetic values, also including considerable historical and symbolic values. This has recently been discussed by Cronon (1995) and by Olwig (1995). One problem is that these ideal landscape types are considered as unchangeable over time; in other words the mental landscape is frozen in time. The dynamic nature of ideal landscapes related to aesthetics has to be stressed. Two areas in particular have been used for research for a long term and they could be used as key reference areas; they are Oxhagen in Scania (Plates 7-9) and Bråbygden in the south-east of Sweden.

5.5 A series of conclusions

Landscape planners are important! They will be needed even more in the future. There will be a need to develop participatory planning, with outside assistance so as to keep a deep and wide knowledge about landscape and landscape management, avoiding simplistic standard methods and clichés. The provision of expert landscape planners is therefore important so that high quality landscape character and local distinctiveness can persist through the development of clear landscape management strategies.

A dynamic time-perspective is essential. It is, however, difficult to achieve when working with landscape, planning and conservation. GIS and other digital visualization methods, and databases of information, may help where previous techniques have failed to work in a dynamic and transparent way. However, the technology of GIS should not replace other people-centred approaches that can be used to develop appropriate strategies.

Landscape laboratories provide an opportunity to identify important elements, habitats and landscape patterns for the future, making it possible to go beyond site specific elements from the past. They give us a chance to test prototypes, in full-scale, in an effective way, by concentrating on specially selected areas, where alternatives can be studied side by side. They form an important supplement to traditional, controlled research stations and to empirical studies out in the landscape.

Reference landscapes are important as a study approach that is concerned with gaining an integrated knowledge of landscape. This essentially balances the knowledge of facts with other aspects that are taken into consideration one by one, but lacking context. The reference landscapes should be identified at different scales; this will facilitate comparisons of different parts of the world, as well as on European, national, regional and local levels.

Approaches and methods developed in the human sciences should be used more frequently so as to make progress within the fields of landscape research, planning and management. They should assist the development of participatory landscape planning, and should be seen as potentially valuable supplements to the methods associated with the natural science.

References

Appleton, J. 1996. *The Experience of Landscape.* Chichester, Wiley.

Bergson, H. 1928. *Den Skapande Utvecklingen.* Stockholm, Wahlström & Widstrand.

Brinckerhoff Jackson, J. 1994. *A Sense of Place, a Sense of Time.* New Haven, Yale University Press.

Bucht, E. 1997. *Public Parks in Sweden 1860-1960. The Planning and Design Discourse.* Doctoral thesis, Swedish University of Agricultural Sciences, Alnarp.

Collingwood, R.G. 1939. *An Autobiography.* Oxford, Oxford University Press.

Cronon, W. 1995. The trouble with wilderness; or getting back to the wrong nature. In Cronon, W. (Ed.) *Uncommon Ground, Towards Reinventing Nature.* New York, Norton & Company. pp. 69-90.

Crowe, N. 1995. *Nature and the Idea of a Man-made World.* Massachusetts, MIT Press.

Flyvbjerg, B. 1991. *Rationalitet og Magt, Bind 1. Det Konkretes Videnskab.* Aalborg, Aalborg Universitet, Akademisk forlag.

Fry, G. and Sarlöv-Herlin, I. 1997. The ecological and amenity functions of woodland edges in the agricultural landscapes; a basis for design and management. *Landscape and Urban Planning*, **37**, 45-55.

Gustavsson, R. 1986. *Struktur i Lövskogslandskap. Structure in the Broadleaved landscape.* Doctoral thesis, Swedish University of Agricultural Sciences, Alnarp.

Gustavsson, R. 1988. Naturskogar i blekinge. *Blekinges Natur, Karlskrona*, (1988), 15-49.

Gustavsson, R. 1991. De försvinnande lundarna. *Blekinges Natur, Karlskrona*, (1991), 57-83.

Gustavsson, R. 1995. *Faculty Research Programme: The Pastoral Landscapes, Perspectives for Landscape Planning, Landscape Management and Grazing.* Alnarp, Department of Landscape Planning, Swedish University of Agricultural Sciences.

Gustavsson, R. 1997. Kunskap med känsla och vördnad. Utomhuspedagogik på Tjärö om landskap och landskapsvård. *Skog & Forskning Skogsvårdsförbundet*, **1/97**. 38-51.

Gustavsson, R. and Fransson, L. 1991. *Furulunds Fure*. Swedish University of Agricultural Sciences, Alnarp, Stad & Land, **96**.

Gustavsson, R. and Ingelög, T. 1994. *Det Nya Landskapet*. Jönköping, Skogsstyrelsens förlag.

Gustavsson, R. and Rizell, M. 1998. *Att anlägga skogsbryn. Concepts and references for designing and reconstructing woodland edges.* Swedish University of Agricultural Sciences, Alnarp, Stad & Land **160**.

Gustavsson, R. and Wittrock, S. 1997. Lunden - om naturen till glädje och nöje. *Skog & Forskning Skogsvårdsförbundet*, **1/97**, 28-37.

Husserl, E. 1995. *Fenomenologins Idé*. Göteborg, Daidalos.

Ihse, M. 1995. Swedish agricultural landscapes - patterns and changes during the last 50 years, studied by aerial photos. *Landscape and Urban Planning*, **31**, 21-37.

Lindquist, B. 1938. *Dalby Söderskog. En Skånsk Lövskog i Forntid och Nutid.* Stockholm, Acta Phytogeographica Suecica, Kartografiska institutet, Esselte.

Olwig, K.R. 1995. Reinventing common nature: Yosemite and Mount Rushmore - a meandering tale of a double nature. In Cronon, W. (Ed.) *Uncommon Ground, Towards Reinventing Nature.* New York, Norton & Company. pp. 379-409.

Peterken, G. 1981. *Woodland Conservation and Management.* London, Chapman and Hall.

Popper, K.R. 1996. *En Värld av Benägenheter.* Stockholm, Brutus Östlings Bokförlag Symposion.

Rackham, O. 1975. *Hayley Wood, its History and Ecology.* Cambridge, Cambridgeshire and Isle of Ely Naturalists´ Trust Ltd.

Rackham, O. 1976. *Trees and Woodland in the British Landscape.* London, Dent.

Sarlöv-Herlin, I. 1991. *Park och natur på Skärva.* Swedish University of Agricultural Sciences, Alnarp, Stad & Land **92**.

Selander, S. 1955. *Det levande landskapet i Sverige.* Stockholm, Bonniers Förlag.

Skånes, H. 1996. *Landscape Changes and Environmental Dynamics.* Doctoral thesis, Stockholm University.

Söndergaard Jensen, F. and Koch, N.E. 1997. Friluftsliv i skovene 1976/77 - 1993/94. *Forskningscentret for Skov & Landskab, Hörsholm*, **20**.

Turner, T. 1996. *City as Landscape, a Post-modern View of Design and Planning.* London, E. & F.N. Spon.

Ward Thompson, C. 1998. Review of research in landscape and woodland perceptions, aesthetics and experience. *Occasional Paper, Landscape Design and Research Unit, Edinburgh College of Art, Edinburgh.*

Whitney, G. 1994. *From Coastal Wilderness to Fruited Plain. A History of Environmental Change in Temperate North America from 1500 to the Present.* Cambridge, Cambridge University Press.

PART TWO

APPLYING CONCEPTS OF LANDSCAPE ASSESSMENT

6 INTEGRATING FOREST DESIGN GUIDANCE AND LANDSCAPE CHARACTER

Simon Bell and Nicholas Shepherd

Summary

1.   Landscape character has been considered an important tool for planning and the design of forests since the 1960s and is incorporated in the Guidelines on Forest and Woodland Design.

2.   Since the 1980s the Forestry Commission has been involved with a number of landscape assessment projects and the development of a methodology. These included the evolution of local design guidelines based on character types.

3.   Other approaches to landscape character assessment based on broader scale methods, such as the ITE land classes, proved promising but were overtaken by the programme developed by SNH, Countryside Commission and Countryside Council for Wales.

4.   Following the preparation of the Dumfries and Galloway Landscape Assessment, the Forestry Commission joined with other partners to develop local forest and woodland design guidance for the region.

5.   Although early indicative forestry strategies disregarded landscape character, it is now included as a major component.

6.1 Introduction

Afforestation, especially using non-native fast growing conifers, has been a major agent of landscape change, especially in many upland areas of Britain. Concern about its effects on the landscape started in the 1930s, became significant in the 1960s and since then has been increasingly addressed in forestry planning and management. While forest managers initially concentrated on the impact of individual afforestation proposals, more emphasis has recently been placed on the impact of forestry on the wider landscape setting. The development and use of landscape character assessment as an aid to this task has progressed spasmodically over the years. Most recently, work based on the landscape character assessment for Dumfries and Galloway has defined a new, proactive direction aimed at guiding how new forestry planting should take place in relation to local landscape considerations.

This chapter reviews the scope, development and applicability of a range of approaches adopted in the search for the best means of addressing landscape character in forest design and management. The most recent Dumfries and Galloway work is demonstrated more fully because it represents the peak of development so far and should point the way forward for any future applications of a similar nature.

6.2 Landscape character

The impact of the new afforestation that commenced shortly after the First World War first showed up when the young trees were some 10 years old. The drive to expand forests to provide a strategic reserve of timber led to a policy where as much land as possible was planted as quickly as possible with the fastest growing and most suitable trees for the sites, which were mainly on upland areas of former rough grazing. Poor soils and a harsh climate led to the use of fast growing evergreen conifers that originally came from places such as North America or Europe. These trees had been collected and tried out on many large estates during the 19th century so were known to be suitable for the new afforestation project. Moreover, they were planted in straight rows in blocks defined by straight 'rides' or compartment boundaries and geometric fence lines. Thus the visual impression from the start was of rectangles of densely planted evergreens laid over the semi-natural vegetation patterns, blanketing any natural features except those few places where it was impossible to plant.

The first concerns about the effect of this new forestry on the landscape were expressed in the 1930s in the Lake District in England, not at that time a National Park, but already a national treasure with its associations with the Lake poets and painters such as Turner. This culminated, in 1936, with an agreement between the Forestry Commission (FC) and the Council for the Protection of Rural England to limit further afforestation and avoid a central core of the Lake District altogether. This agreement resolved the problem by zoning, not by planning or design. At that time ideas about fitting forests into the landscape were not considered and would not have been acceptable to most foresters.

After being interrupted by the Second World War, planting recommenced at an increased rate. Simultaneously, especially in the 1950s and 1960s, the countryside became an increasingly popular destination for recreation. Many of the forests planted in the 1920s and 1930s were opened for public use and the issue of landscape impact of forests moved up the agenda. In 1963 Miss (later Dame) Sylvia Crowe, an eminent landscape architect who had worked on many major public projects such as motorways, reservoirs and power stations, was appointed as the Forestry Commission's first landscape consultant. She had to start from scratch in developing ways of accommodating new forests into landscapes, particularly the most sensitive. Sylvia Crowe had to develop the first set of design principles in order to be able to demonstrate to sceptical foresters that forests could be made to look better and be practical at the same time. She won the respect of many foresters and laid the foundations for the development of forest design as a new discipline, now carried on by a number of landscape architects employed by the Forestry Commission.

Sylvia Crowe started by analysing the character of the landscape within which each new afforestation project was to fit. Often constrained by ownership boundaries, she tried to reflect the following criteria (Crowe, 1978).

- Follow the landform (often a major influence in the uplands) and avoid placing geometric shapes that cut across it at awkward angles.
- Borrow from vegetation patterns (sometimes foresters already followed vegetation with the tree species they planted).
- Increase diversity by incorporating open spaces, tree species variety, native broadleaves and feathered edges.
- Tie the forest into the landscape by using hedgerows, stream valleys, rock outcrops and other natural or cultural elements.

Because afforestation was dependent on land availability, its progress was an *ad hoc* affair and never the subject of a land use strategy. This continues to the present day, although documents such as Indicative Forestry Strategies attempt to give some kind of direction (see section 6.4).

From the initial work by Sylvia Crowe, the comprehensive range of forest design principles has been developed. These now apply also to felling and replanting, where there is a major opportunity to improve the design of the earlier forests. Perhaps the main criticism that can be levelled at the forestry industry is that, despite the well developed principles and methods of applying them, the rate of progress of improvement to forest planting was not as great as it could have been. This reflected policy priorities and perceptions of landscape sensitivity. However, each successive development of instruments, such as the Woodland Grant Scheme, enables better quality to be achieved.

The main vehicle for ensuring that good design is accommodated is the suite of design guidelines, commencing in 1989 with the *Forest Landscape Design Guidelines* (Forestry Commission, 1989), supported in 1991 by *The Design of Forest Landscapes* (Lucas, 1991). These books not only contained the standard to be aimed for, but also the principles and their application. Central to that is the need to assess landscape character, though this integral part of the design process tends to be restricted to the immediate setting of the particular design project. Such broad brush guidance may lead, if not adequately informed by a rigorous analysis of the local landscape character, to somewhat 'cook book' solutions. *The UK Forestry Standard*, published in 1998, seeks to avoid standardisation. This is where localised guidance based on more extensive landscape assessments is likely to have the greatest potential.

The use of landscape assessment and attempts to incorporate it into forest design have a chequered history. The first ways of assessing landscapes in general tried, for example, to measure qualities or values using numeric scores for different features related to kilometre squares (Fines, 1968). These attempted to make objective the subjective values and to rely on expert judgement to set criteria in the first place. Such measures of value have been used to establish which landscapes should be protected, but have little use for guiding how landscape change can be accommodated. Landscape value has been incorporated into Indicative Forestry Strategies; this is one of their shortcomings (see section 6.4).

The idea of describing or analysing landscape character as a separate task from ascribing value was reawakened in the 1980s. A number of methods developed, some more subjective than others. A principal concern, not yet fully addressed in any method, has been how to describe visual or aesthetic features in rational or objective terms. Currently the best methods are strongest at describing land character rather than landscape character, although the division is by no means a hard one. Land character is the synthesis of the physical, ecological and cultural factors that have made the landscape what it is, whereas landscape character incorporates human perceptions. The difficulties are 'whose perceptions?' and how to gauge them. Various research projects have attempted, with some success, to find out more about public perceptions of forest landscape (Lee, in press; Bell, 1998a).

6.3 The development of landscape assessment

The Warwickshire Landscape Assessment study was one of the first that started the current trend of landscape assessment using the approach of landscape description and analysis, building up from layers of geology, landform, soil, settlement, land use, archaeology, cultural history and ecology. This analysis was conceived at two levels, regional and local. Regional landscape areas extend well beyond the boundaries of Warwickshire, reflecting the main zones of geology and land use in the English Midlands. Within each regional division there can be a number of local landscape character types.

The method of creating this regional assessment was extremely objective. A TWINSPAN analysis was used to sample a number of grid squares for attributes relating to biophysical, land use and other cultural factors and ascribe these squares into a number of classes (Warnock and Cooper, no date). Thus, discrete units of distinct landscape character emerge from this approach. The description of landscape character is different from the measurements of landscape value alluded to earlier, and much more useful for planning and design purposes (the landscape can nevertheless be evaluated for purposes of protection as a separate exercise). The landscape perceptive element remains a weakness in this method.

Once the basic character has been established, the current condition and tendency for change of that landscape can also be incorporated and used as the basis for countryside management strategies, including woodland extension and management. Thus a landscape strategy with accompanying design guidelines was produced for the whole of the county (Warwickshire County Council, 1991).

The 'Warwickshire Method' has been used extensively with various modifications, to suit different circumstances, and has proven to be very robust. Its development was supported by the Countryside Commission. The Forestry Commission also took an active interest in the project. On the basis of the results achieved for Warwickshire, a more forestry-directed approach by Staffordshire County Council was provided with direct support in staff time.

One of the first English counties to embark on the preparation of an Indicative Forestry Strategy (IFS) (see section 6.4) was Staffordshire. The Forestry Commission offered to assist by assigning a member of staff to the team assembled by the county. The result was a two-level landscape assessment, of regional types and local types similar to that of Warwickshire and produced by almost identical methods including the TWINSPAN analysis. For each type a detail sheet was prepared; this applied design principles and suggested how new woodlands might be accommodated, for example by considering their position, shape, scale and species. A publication describing how to apply the method of assessment used in Staffordshire as part of the preparation of the IFS was produced by the Forestry Authority in England (Price, 1993).

In Wales a similar route was followed, where the Forestry Commission joined forces with Clwyd County Council to prepare a landscape assessment and an IFS. This assessment followed the Warwickshire and Staffordshire methods. One of the drawbacks of these for use in the late 1990s is their laborious manual preparation. The increased availability of geographic information systems should help in speeding up the process and aiding management and retrieval of the information, although the actual analysis and synthesis into landscape character areas is less easily achieved by using these systems.

Subsequently, the development and application of landscape assessment in Wales has pursued a different route, using a system called 'Landmap', that is based around a multi-layered GIS. However, this has evolved into more of a land planning and management tool. Currently it does not fully incorporate the synthesis into landscape character as understood here. Landscape zones are incorporated as a layer but they risk becoming areas of value rather than character. Notwithstanding these reservations, Landmap is likely to become a valuable land information resource with the capacity to aid a number of land use strategies in Wales.

Other examples of the use of landscape character assessment used to drive forestry and woodland strategies and to supply the basis for advice can be found in the 12 Community Forests and the National Forest in England. These landscapes are often considered to be degraded, so the assessments are used but only to identify areas where the existing character is well defined and robust, but also, and more importantly, where a new character can be created. This proactive side is often overlooked in the concerns for landscape protection or conservation.

One of the issues arising from these county based projects is the amount of work involved and the sheer scale of applying such detail all over the country, especially in the lowlands. At the same time that these studies were taking place some other possible approaches for achieving less detailed but still localised results were explored.

The question arose 'how many local design guidelines would be needed for the whole country?'. Could it be the case that there might only be a limited number because similar landscape types might occur on similar geologies or with similar land uses in different places? If this was the case, then it would make the provision of guidance much simpler.

The Institute of Terrestrial Ecology (ITE) at Merlewood has developed the land classification system (Bunce *et al.*, 1981). This uses samples taken from across the country and classifies them into a limited number of classes of land. The system samples a number of variables including geology, soils, landform, land use, building types, and so on. A booklet was produced that illustrated the classes, showed where they occurred and described their characteristics (Benefield and Bunce, 1982).

One of the advantages of the land classification system is that it presents an objective method and, because the samples are reassessed, it is periodically updated. In order to be as rigorous in the landscape description as in the classification, the visual vocabulary of design principles developed as a means of conveying concepts of the visual landscape to non-designers such as foresters (Bell, 1993). This relies on expert judgement but remains descriptive and implies no value. Moreover, since a number of the design principles have now been tested against public preferences (Lee, in press; Bell, 1998b), a relationship has been established between expert description using the design vocabulary, the application of the principles to design, and the preferences of the public for the design prepared in this way.

Some of the most widely occurring lowland land classes were used as a basis for examples of how to analyse and design new woodlands presented in the Lowland Landscape Design Guidelines (Forestry Commission, 1992). From this, a number of basic landscape types, also based in part on the ITE land classes, were used in research into preferences for woodland design (Bell, 1998b). However, other developments elsewhere, such as the programme of Scottish landscape character assessments and projects such as the Countryside Character Programme and Landmap have displaced the use of land classes in this field. This may be a missed opportunity because the objective and nationwide coverages of land classes could have provided a very useful base on which to develop a wide range of guidance. The drawback of the land classes is their origination from a limited number of samples so that the local nuances are missed. However, as a regional framework within which to locate and help guide the development of the local assessments, they may still have a value.

Concurrent with the Forestry Commission's land class based project, the Countryside Commission embarked on the 'New Map of England'. This was a pilot regional landscape character assessment project also using the TWINSPAN computer analysis to generate the landscape types and then to use descriptive techniques and public involvement as a means of defining their individual properties. The landscape character types generated from this were varied in extent and could be quite clearly identified on the ground. This map has subsequently been further developed by incorporating English Nature's Natural Areas programme into the Countryside Character Programme (Countryside Commission and English Nature, 1996). The development of forestry guidance on any of these bases has not yet been carried forward because of the slow progress of woodland planting in England over recent years. However, it may be of more use in the future following the development of the *English Forestry Strategy* (Forestry Commission, 1998). The most recent, and by far the most comprehensive development has been to prepare local design guidance for Dumfries and Galloway (see section 6.5).

6.4 Indicative Forestry Strategies

In section 6.2 the lack of forest or land use strategies in GB was noted. The nearest approximation to forestry strategies are Indicative Forestry Strategies (IFSs). These aim to demonstrate where the accumulation of environmental sensitivities may constrain afforestation to a greater or lesser degree.

The first IFS was prepared by Strathclyde Regional Council (Strathclyde Regional Council, 1988). It was essentially produced using a map layering exercise, starting with the physical limitations to afforestation and overlaying this with different sensitivities, of which landscape (or scenery) was one. Landscape sensitivity was based mainly on existing designations such as National Scenic Areas or local Areas of Great Landscape Value. The IFS was expressed as a map of zones showing sensitive, potential or preferred areas, together with places where afforestation was physically or climatically excluded, thus giving some guidance as to where the council would prefer to see more or less afforestation.

These early IFSs thus assumed that sensitivity to afforestation was greater where scenic quality was deemed to be higher. There was no attempt to consider how areas of different landscape character could accommodate forestry in different ways. For example, some highly attractive landscapes with diverse vegetation and complex landforms can easily accept forestry and are easier to design into than other, blander and less diverse, though less sensitive, landscapes. This is a shortcoming of IFSs that could be rectified by revising them to incorporate either a landscape character assessment, local woodland design guidelines, or both.

In the 1990s the idea of extending the application of IFSs into England and Wales was followed up with a joint Department of the Environment and Welsh Office planning circular on the preparation of IFSs (Department of the Environment and Welsh Office, 1992). While the circular was largely based on the Scottish models, landscape character was given a larger part. This was because, when compared with the Scottish examples, there were likely to be fewer large scale nature conservation constraint areas, fewer physically or climatically restricted areas, except urban and derelict land, and, in lowland areas, fewer obvious areas carrying high scenic values. Thus the scope for afforestation largely depended on the 'carrying capacity' of the landscape, which in turn hinges on its character.

6.5 Landscape design guidance for forests and woodlands in Dumfries and Galloway

The Forestry Commission has been a consistent supporter of the SNH Landscape Character Assessment Programme since its inception. As consultee, FC have reviewed and commented on a high proportion of the draft assessments, especially with respect to forest and woodland issues. FC did, however, become more involved with the process of developing one particular landscape character assessment. For the Dumfries and Galloway Landscape Assessment (Land Use Consultants, 1998), FC accepted an invitation to become an active partner on the project Steering Group. From that involvement arose the conviction that there could be significant benefit for all those involved in the establishment and management of forests and woodlands to extend and expand on the advice on those topics within the landscape assessment that was envisaged.

The original concept was for forest and woodland design Guidance Sheets to provide basic technical advice on forest landscape design, related to specific landscape character areas. Fundamentally, a marriage between the general FC *Forest Landscape Design Guidelines* (Forestry Commission, 1989) and the specific recommendations of the landscape assessment.

The project was developed and funded by three of the four original partners for the landscape assessment, including Dumfries and Galloway Council, Scottish Natural Heritage, and Forestry Commission. Environmental Resource Management (ERM) were the consultants commissioned to carry out the study in February 1997. Although the contract has been completed, the results will be subject to extensive consultation, probably starting in Spring 1999. The consultant's brief outlined two main objectives of the project. First, to build on and expand the advice on forest and woodland development and management in the landscape assessment. Secondly, to demonstrate the principles of forest and woodland design advocated in the FC design guidelines for the landscape character types of Dumfries and Galloway.

This design guidance was intended to be of interest generally to those involved in the design and management of forests and woodlands, and specifically to those seeking FC grant aid for their proposals through the Woodland Grant Scheme. It would also, however, be of interest to the Dumfries and Galloway Council, non-governmental organisations and community groups involved in the consultation process for forest and woodland proposals throughout Dumfries and Galloway.

The benefits of the design guidance to forest and woodland managers was considered twofold. First, it would assist with the development of proposals appropriate to the opportunities and constraints of the landscape and the site. Secondly, it should ensure rapid consultation and approval of proposals, recognising that the design guidance has been developed with the consensus approval of the primary consultees. The consultees themselves would also benefit from the advice, being better informed in their appraisal of forest and woodland design and management proposals throughout the landscape character types of the area.

Each Guidance Sheet covering a specific landscape character type (or, in some cases, two or three similar types) has been developed to provide a consistent level of information, all laid out to a standard format (see Plates 11 and 12). Each describes how landscape change can best be accommodated and what positive measures could be taken to conserve, improve or restructure an afforested or woodland landscape.

The standard format is for the front of each sheet to be an appraisal of the existing landscape character. This includes some details of local characteristics, heritage features and a summary of the main ecological influences. These characteristics are also illustrated by photographs of typical scenes and a composite sketch of the landscape type. The reverse side of each sheet provides succinct design guidance, describing how new woodland and forests could be integrated with the existing landscape character. This advice is supported by sketches illustrating the potential visual impact of suggested changes to the woodland cover of the area. The Guidance Sheets are intended to be available on a 'pick and mix' basis, held within a binder. Each binder will include a comprehensive introduction, location map of each landscape character type and a glossary of terms used.

6.6 Conclusions

Landscape has assumed a growing importance as one of the most significant environmental and social issues to be considered in forest planning in Great Britain. The majority of landscapes where more forest planting might take place already possess a strong character, the result of many processes, both natural and cultural. Thus, it is important to ensure that this landscape legacy is not compromised, while allowing landscape change to continue as it always has, albeit with a more defined direction.

At present there are no plans to expand the programme of design guideline development for local areas on the basis of the landscape character assessments. However, the assessments themselves, the Countryside Character Programme in England and the Landmap process in Wales will certainly be used to provide valuable context and background of a strategic nature. The latest development, that of Forestry Frameworks for parts of Scotland, is expected to make use of the relevant landscape character assessments. Experience of the various projects in which the Forestry Commission has been involved suggest that the following four points are important to ensure their value.

- Landscape character assessments should be based, as far as possible, on objectivity in their description and analysis. They should also be consistently applied so that as they build up across a country they fit seamlessly together.
- Description and characterisation should be separate steps from evaluation. Some agencies are more interested in the latter than others. Evaluation of quality and sensitivity are useful for some strategic purposes, especially in terms of where to target resources.
- Where landscape assessments are used for preparing local design guidelines of whatever variety (not only forestry and woodland), local communities and land owning interests must be involved. Land owners especially must not see design guidance as prescriptive in any way. Illustrations used to demonstrate applications must show anonymous yet typical examples.
- More effort is needed to incorporate an established visual vocabulary as part of a more consistent and rational character description.

As the political landscape in Scotland and Wales changes following devolution, it is likely that more regional strategies for forestry will develop. Landscape character, if used following the four points described above, should become an important aspect to be included in the development of such strategies.

References

Bell, S. (1993). *Elements of Visual Design in the Landscape.* London, E&FN Spon.

Bell, S. (1998a). Woodlands in the Landscape. In Atherden, M.A. and Butlin, R.A. (Eds.). *Woodland in the Landscape: Past and Future Perspectives.* York, The Place Centre, 168-182.

Bell, S. (1998b). *The Landscape Value of Farm Woods.* Forestry Commission Information Note 13.

Benefield, C.E. and Bunce, R.G.H. (1982). *A Preliminary Visual Presentation of Land Classes in Britain.* Grange-over Sands, Institute of Terrestrial Ecology.

Bunce, R.G.H., Barr, C.J. and Whittaker, H.A. (1981). *Land Classes in Great Britain: Preliminary Descriptions for Users of the Merlewood Method of Land Classification.* Grange-over-Sands, Institute of Terrestrial Ecology.

Countryside Commission and English Nature (1996). *The Character of England.* Cheltenham, Countryside Commission.

Crowe, S. (1978). *The Landscape of Forests and Woods.* Forestry Commission Booklet 44.

Department of the Environment and Welsh Office (1992). *Indicative Forestry Strategies.* Joint Circulars 29/92 and 61/92 respectively. London, HMSO.

Fines, K.D. (1968). Landscape evaluation: a research project in East Sussex. *Regional Studies,* **2,** 41-55.

Forestry Commission (1989). *Forest Landscape Design Guidelines.* London, HMSO.

Forestry Commission (1992). *Lowland Landscape Design Guidelines.* London, HMSO.

Forestry Commission. (1998). *England Forestry Strategy: a New Focus for England's Woodlands: Strategic Priorities and Programmes.* Cambridge, Forestry Commission.

Land Use Consultants (1998). *Dumfries and Galloway Landscape Assessment.* Scottish Natural Heritage Review No. 94.

Lee, T.R. (in press). *Forests, Woods and People's Preferences.* Forestry Commission Technical Paper 18.

Lucas, O.W.R. (1991). *The Design of Forest Landscapes.* Oxford, Oxford University Press.

Price, G. (1993). *Landscape Assessment for Indicative Forestry Strategies.* Cambridge, Forestry Authority..

Strathclyde Regional Council (1988). *Forestry Strategy for Strathclyde.* Glasgow, Strathclyde Regional Council.

Warnock, S. and Cooper, A. (No date). A Regional Landscape Classification for the Midlands. Unpublished report to the Countryside Commission.

Warwickshire County Council (1991). *Arden Landscape Guidelines.* Warwick, Warwickshire County Council.

7 SUSTAINABLE TOURISM AND THE LANDSCAPE RESOURCE: A SENSE OF PLACE

Duncan Bryden

Summary

1. The perception of landscape character by visitors is fundamental to tourism in Scotland.
2. Tourism is a high value industry, showing signs of steady long-term growth. There is evidence that certain market segments exhibit preferences for particular landscape types.
3. It is difficult to make direct links between landscape value and economic value.
4. Landscape planning processes and designations could integrate the interests of the tourism industry, and particular niche segments, better. The tourism industry needs to support good planning and management practice to minimise impacts and to make better use of the landscape potential.
5. Area Tourism Strategies are one vehicle for enhanced integration between landscape issues and the tourism management process.

7.1 Introduction

Landscape character has a critical role to play in attracting the visitor to Scotland. The visitor may call it scenery or views and they may find it difficult to articulate the complex interaction of characteristics that make up a scene or a view, but visitors do exhibit preferences. Each individual views a landscape in a different way and has personal choices and values. Yet some correlations occur across demographic types in relation to certain landscapes, particularly mountains and coasts, which allow them to be further described in terms such as impressive or dramatic and to be singled out by the visitor as a significant reason to visit (Anon., 1997b).

Landscape is so important to tourism that we need to know more about visitor landscape preferences and match these more carefully with the market driven approach of the tourism industry. The marketing effort by the Scottish Tourist Board is focused around a small number of popular landscape types and although this reinforces existing images, and will influence visitors to come to Scotland, it does not always reflect the diversity of Scottish landscapes. Areas without dramatic mountain and coastal landscapes are disadvantaged in attracting visitors and their own local distinctiveness can be neglected. This chapter contends that if the Scottish tourism industry is to become more sustainable it should play a greater role in framing the objectives and policies which will influence landscape character and the tourism industry should reassess its own approach to marketing landscapes.

The chapter looks at four factors that need to be addressed if the tourism industry is to make better use of the landscape resource. First, there is a need to define the scope and extent of Scottish tourism and what part landscape plays in sustainable tourism. Second, it considers the existing relationship between Scottish tourism and the landscape. Third, it addresses the links between landscape planning and the tourism industry. Fourth, it explores the actions that could increase the role of the tourism industry in influencing landscape change.

7.2 Landscape and Scottish tourism

Before considering Scottish tourism it is important to appreciate the significance of tourism worldwide. The word 'tourism' was recorded by the Oxford English Dictionary as appearing for the first time in 1811 when castles, churches and other cultural artefacts began to experience the tourist's gaze. Apart from pilgrimages, tourism really began when the Grand Tour became established in the eighteenth century and heralded greater numbers travelling for pleasure, education and knowledge. Tourism has developed from that time to the situation where international arrivals grew from 25.3 million per annum in 1950 to a projected 661 million per annum in the year 2000.

> *" Today, travel and tourism is the world's largest industry, transporting more than 528 million people internationally and generating $322,000 million in receipts in 1994. It is a major economic force, generating in 1995 an estimated $3.4 trillion in gross output creating employment for 211.7 million people and producing 10.9 per cent of the world's gross domestic product (GDP)."* (Anon., 1996, pp. 33-34).

The industry is becoming much more segmented by country of origin and by activity. Scotland's contribution from tourism is in the order of £2,500 million, creating employment for 177,000 people and 5 per cent of GDP (Anon., 1997c). Tourism's contribution is over £500 million more than agriculture and is on a par with whisky exports (Scottish Enterprise, pers. comm.). With this position of influence on the economy, it should have financial self-interest at heart, for the tourism product that Scotland packages and sells ultimately relies on a quality environment including clean water, litter free streets, well conserved habitats and landscapes, and a vibrant culture.

Scottish tourism cannot become more sustainable without the industry playing a greater role in the future management which maintains the quality of the diverse landscapes that constitute Scotland's elemental character. Equally, tourism has to ensure its own standards of infrastructure and visitor management. The industry recognises the role that quality landscapes play in meeting the expectations of visitors to Scotland. Middleton and Hawkins (1998, p. ix) described sustainable tourism as having " a particular combination of numbers and types of visitors, the cumulative effect of whose activities at a given destination, together with the actions of the servicing businesses, can continue into the foreseeable future without damaging the quality of the environment on which the activities are based."

For practical decisions in tourism, environment means the quality of natural resources such as landscape, air, water and ecological components. Landscapes in Scotland also contain significant cultural and even emotional elements which contribute to the local distinctiveness in tandem with natural heritage values. According to Baxter and Thompson (1995, p. 48) " there are scarcely enough places anywhere in Britain without the hand of man".

7.3 Existing relationships between Scottish tourism and the landscape

Landscapes, particularly of the natural type, contribute directly to the sense of place experienced by a visitor to a destination. Scotland's image has been honed by the pen, paintbrush and camera of travellers over many centuries, shaping the landscape preferences of future tourists. Well documented accounts including Hunter (1995) and others refer to the writings of tourists travelling in Scotland in the eighteenth and nineteenth centuries. Romantics including William and Dorothy Wordsworth, Sir Walter Scott and before them James MacPherson, engaged in communicating almost fantastical descriptions and illustrations of the Scottish landscape. Doctor Samuel Johnson and James Boswell are at least credited with reporting life as it was rather than the romantic idea of what it should be. Visiting waterfalls, lochs and ruined castles was the essence of Victorian tourism in Scotland. This romantic tradition was retained at least up to the 1930s by writers such as Ratcliffe Barnett and Alistair Alpin MacGregor, describing engaging landscapes populated by a couthy local culture. The Great North Road from Perth described by Findlay (1976) was completed by 1935 to an acceptable standard for private cars, providing ready personal access to much more of Scotland. Post war a more realistic type of guidebook emerged, but much of the romance created by the early writers still remains part of modern tourism marketing in Scotland.

Scottish Tourist Board research records that over 80 per cent of visitors consider scenery the most 'liked' feature of their visit to Scotland and a similar figure is recorded for individual destinations such as Loch Lomond. Use of natural heritage imagery by the Scottish Tourist Board (STB) and the Area Tourist Boards (ATBs) clearly reflects continuing customer enthusiasm and appreciation for rugged countryside.

> " The best reason for choosing to go on holiday to Scotland is this: it is one of the last places inside the crowded and frenetic European Community where it is possible, indeed easy, to be alone in empty countryside" (Anon., 1997c).

Guidebooks are an increasingly important way of finding out about Scotland. The 1997 Highland Visitor Survey recorded that 13 per cent of all visitors were influenced by guidebooks; tourist brochures, traditionally the main marketing vehicle, also recorded 13 per cent but are losing ground (Anon., 1997b). Guide book authors recognise the importance of the landscape in marketing Scotland.

> " Scotland remains, as we have stressed all along, one of the best countries in which to unwind in the middle of landscape that can be matched nowhere else in these islands" (Leslie, 1997, p. 23).

> " Forget the castles, forget the towns and villages. The spectacular Highlands are all about mountains, sea, heather, moors, lochs and wide empty exhilarating space" (Smallman and Cornwallis, 1999, p. 309).

What is significant and important for visitors is that the Scottish landscape produces a distinctive scene to be enjoyed, an experience often far removed from that which they encounter in everyday life. The visitor must experience something out of the ordinary according to Urry (1990). Marketers have translated this landscape experience into a unique selling point for Scotland. Typical campaigns on the London Underground convey images of unique wild landscapes to stressed commuters. Pithy captions including 'leaves you breathless, like the air in London' reinforce the images.

Tourism has been linked with concerns over the serious deterioration of the natural environment in parts of Scotland. However, following extensive consultation, a Scottish Tourism Co-ordination Group report in 1992 concluded that although impacts from tourism were wide spread, there were only serious problems in a few localised areas. MacLellan (1998) pointed to other industries having a much more significant effect on the environment and the impacts of tourism should be kept in perspective. He concluded that the over-reaction could be attributed to tourism's high media profile and a resistance from supporters of traditional industries. This criticism is explained further by Farrell and MacLellan (1997): A high proportion of well–educated intellectuals, through possibly ignorance or emotion, attribute an unreasonable amount of responsibility for landscape change and environmental degradation to tourism.

7.4 Links between landscape planning and the tourism industry

Increasingly sophisticated and demanding consumers are concerned with the way that the landscape asset is treated and both the product and the image must be enhanced to maintain the tourism industry's competitiveness. Landscapes, both extensive and intimate, are a vital, if often a subconscious, backdrop to tourism operations and a fundamental part of the marketing effort. The natural environment is both Scotland's top tourism product and its brand image (Scottish Enterprise, pers. comm.).

Landscape images and descriptions in tourist brochures, guidebooks, posters, paintings, films and television advertising form a primary perceptual filter as visitors make key decisions whether to even consider visiting the place. Wild, open landscapes dominate the brochures and serve as a scenic backdrop for a wide range of activities and outdoor pursuits. Entire marketing campaigns such as 'Autumn Gold' draw their name from landscape imagery. The images illustrating Scottish villages and towns are of those scenes that show buildings made attractive by their traditional nature through scale, texture, colour, roof pitch, window size, building materials and surrounding streetscapes. Scottish farming and countryside schemes are depicted by classic, almost old fashioned, types of agriculture: free range livestock contained within drystone wall and hedged field boundaries and small scale farming. Large scale cereal farms, intensive livestock units or monocultural Sitka spruce plantations tend to not feature in tourist brochures. Images of forests and woodlands show more natural looking design and autumn colours. Cityscape photographs are shot from a distance or in flattering light and unattractive suburbs or industrial complexes never appear. Marketers value such images to the extent that they will 'doctor' photographs. Powerlines, foreground rubbish and other undesirable elements can be 'removed' in order to get the ideal, desired landscape if it does not conveniently exist.

There are certainly market segments attracted by particular landscape attributes and features. Visitors from the larger European countries, such as Germany and France, considered great stretches of wild country as being particularly attractive. The Dutch value particular aspects of the landscape. 'Hills, mountains and cliffs are unknown for Dutch people' (MacPherson Research, 1998). Many mainland European visitors seemed attracted by the 'accessibility' of the landscape, both in terms of footpaths and also with the openness arising from a lack of field boundaries and woodlands, as well as the ease with which it is possible to interact with the landscape. These are features that are fundamental to the choices made by visitors in particular market segments before visiting an area.

Amongst those who seek to interact more intimately with the landscape, views appear to be particularly important. Herries (1998) in the Glen Shiel Hillwalking Survey stated that 37 per cent of respondents specified that views gave them most enjoyment and a further 25 per cent of positive responses could be classed as view related. Amongst the responses there was little indication of what constituted attractive views, although visitors appeared disappointed by smaller scale features including poor path quality, litter and erosion.

These are the more obvious landscape links with particular market segments; yet much of Scotland has neither wildland nor dramatic mountains or coastline. Denmark has a very intense land use regime and, as in Scotland, relies heavily on the German market for its tourism. It has identified and conserved its particular landscape strengths of sheltered coastlines and low lying cycle-friendly terrain (Scottish Tourism Research Unit, pers. comm.). In Scotland the tourism industry has more work to do in better identification of the more subtle needs of market segments in relation to the local landscape. There is a requirement for more research from both landscape professionals and the tourism industry to identify visitor needs. There is considerable literature on landscape evaluation research (see particularly Hughes and Buchan, this volume), although the values and interests of a professional elite need to be reconciled with those of the general public. This will be difficult and increasingly complex as market segments and associated visitor preferences for landscapes and associated activities, infrastructure and transport requirements can change over time.

Walking, riding, fishing and driving are traditional activities that are greatly enhanced if they are practised in attractive areas even if the value of that aesthetic factor is difficult to calculate in financial terms. Visitors enjoying new outdoor pursuits from paragliding to mountain biking also prefer to use attractive areas for their activities. Other market segments such as those engaged in windsurfing or 'cycling for softies' may require a very different landscape resource for their recreation. Uneven benefits occur where 'improvements' for one tourism activity, for example field sports, can cause major landscape changes. Bulldozed access roads on hills and croys in rivers can impact to the visual detriment of many other landscape users.

Tourism is a very mobile industry. New destinations and market segments can appear very rapidly and existing destinations can lose their appeal to certain market segments. An example of where tourism infrastructure becomes out of step with the surrounding landscape is in the downward spiral of the Scottish seaside resort. The descent was accelerated by the appearance of cheap mass air travel, allowing many more to afford a seaside holiday in a warm sunny climate. Scottish seaside resorts and their coastal landscapes lost their integrated approach and became subjected to brash new developments, as caravan sites and amusement arcades vied for a different market groups. The Aviemore Centre exhibited a similar trend as it moved from a vibrant development phase in the late

1960s into a 'make do and mend' mentality from which it has never recovered. The landscape surrounding Aviemore is still there, but its character was not reflected in local developments which in turn hindered visitor access to the adjacent National Nature Reserves and River Spey. Lessons have been learned the hard way. Local and more extensive landscapes can change over decades and visitors will depart. Murray (1962, p.19) writing prior to major tourism growth, intimated that 'the outstanding beauty of the Highland scene, which is one of the nation's great natural assets, has been haphazardly expended and no account kept.'

Tourism interests in the transport network have concentrated on effective dispersal in terms of time. Perhaps less attention has been given to the quality of the trip. The private car accounts for 67 per cent of tourist trips. Leisure trips account for 45 per cent of all journeys and this sector has made the largest contribution to the growth in car journeys in recent years (Anon., 1995a; Scottish Enterprise, pers. comm.). Rural roads are part and parcel of the tourism product. The preferences of visitors and the economic value of tourism are not fully considered by professionals in road design and layout. Few visitors drive through Scotland trying to get from A to B as quickly as possible. Quiet back roads through attractive countryside are an important part of Scotland's visitor appeal. The cumulative impacts of road improvements on some of these routes may affect tourist perception of an area, particularly the tranquillity and naturalness of the area. Road improvements, inappropriate signs, rumble strips, road marking and sterile verges may change the character of a village.

National Tourist Routes (NTRs) have been set up to provide alternatives to motorways and busy through routes and will undoubtedly attract visitors; but signs alone are not sufficient. NTRs do not guarantee scenic quality or any agreed scheme of management along the road corridor (Innes, 1998; Scottish Enterprise, pers. comm.)

Wider, faster trunk roads will result in traditional bridges being replaced with modern structures, loss of hedgerows and roadside trees, deeper road cuttings, night glow, exhaust fumes and a wider impact beyond the road corridor (Anon., 1997d). Road construction regimes that leave a legacy of grass cutting, herbicide application and non-native species planted in regimented rows are expensive and will reduce biodiversity. Good practice guides exist for trunk roads (Anon., 1998b) and some roads like the A9 present attractive vistas although much more remains to be done to provide non commercial, attractive rest areas for the visitor away from the traffic. Certain Area Tourist Boards have expressed concern over 'the lack of a national overview' regarding telecommunication masts along scenic routes and wind farms in environmentally sensitive areas (Argyll, the Islands, Loch Lomond, Stirling and the Trossachs Tourist Board, pers. comm.; Anon., 1997d) and have requested that legislation be tightened up.

Visitor preferences are not being taken sufficiently into account in landscape management. Landscape designations such as National Scenic Areas (NSAs) or Environmentally Sensitive Areas (ESAs) and other protected areas leading to landscape conservation are of little consequence to the visitor. At best they are bewildered at the variety; at worst they are unaware of the designations or their significance. Tourism literature tends to ignore such designations and there is rarely an indication on the ground that informs the visitor they are about to enter an NSA or an ESA. Communities too appear to show little concern for such designations and all too often see them as impediments to development rather than as accolades. In the absence of good research planners are unable objectively to take account of visitor preferences.

Primary industries can have a major impact on the landscape. It is virtually impossible to measure how many forest clear fell areas, rock quarries, power lines or industrial estates active in a landscape might eventually cause visitors to stay away. Market research may give an indication of visitor preferences, but to date this has provided little part of the justification for landscape protection. Local people can be in favour of primary industries, particularly if they generate employment. According to a survey by the operators of the Novar Wind Farm near Alness only 3 per cent of local people expressed opposition to the project and no-one believed visitor numbers had been affected, even though the wind turbines are visible from a number of prominent locations, including a popular tourist route (Anon., 1998c). Following the public enquiry on the Flowerdale hydro scheme in Wester Ross the reporter to the planning enquiry recognised the impact that the scheme would have on visitor enjoyment of the area, but acknowledged that impact would be difficult to measure!

Visitor perceptions of particular landscape features are perhaps easier to record. Gourlay and Slee (1998) considered public preference for certain landscape features supported by ESA funding. They noted that many of the features promoted under the ESA policy appear to be in accord with public preference. Litter free areas, promoting wild flowers, broad-leaved woodland and footpaths would appear to have particularly widespread support. But they concluded that there were some doubts as to the value visitors place on drystone walls (the object of most ESA funding) as landscape features in comparison to other aspects of the ESA landscape. Drystone walls were more popular with residents. The planned 1992/93 expenditure in the Loch Lomond ESA on drystone walls was almost £200,000. This additional sum is equivalent to approximately 20 per cent of the regional park core budget, which had to provide for 5 million visitors. The level of ESA expenditure raises some concerns that landscape benefits delivered by the drystone walls to tax paying visitors are not being adequately realised and that the expenditure might give a greater return in some other landscape measure. Given these beliefs and concerns, it was concluded that more research was necessary to quantify economic links between landscape features and the tourism industry.

If engaging the operator, the visitor and the community in landscape decisions is difficult, it is equally problematic to assess the capacity of landscapes to absorb visitors and their facilities. The figure of 80 per cent of visitors enjoying the scenery as the feature they most liked on their visit cannot yet be readily broken down into its constituent parts. What aspects of the wider scenery visitors like, and indeed what aspects they do not like, remain insufficiently clear to drive policy. Limited information is beginning to emerge. In the Highlands Visitor Survey in 1997, when asked as to their main reason for visiting, 22 per cent indicated scenery and scenic views and 12 per cent specified mountain and hill landscapes. There are clearly also specific segments of visitors in Scotland who do more than just gaze and have an active involvement in the landscape. Over 700,000 climbers and walkers visit the high hills annually (Highlands and Islands Enterprise, pers. comm.). This points to the need for planners to better engage the visitor in quantifiable landscape market research.

Landscapes have a physical visitor capacity, which if exceeded can harm the visitor experience. Physical damage to landscape character by visitors can be locally severe. Linear erosion along tracks and footpaths and damage to sensitive habitats on water margins and sand dunes, or a machair, are common at certain visitor hotspots. But given sufficient funds restoration work at these sites is possible, although it is important that the restoration does not cause further or greater landscape impact. Appropriate measures deemed necessary to manage visitors can include car parks, interpretation, signage, hardened footpaths and visitor facilities. Physical settings devoted to servicing visitors can also be seriously compromised as described by Hunter (1995, p. 11) " Tyndrum, its indigenous architecture long since overwhelmed by as tasteless a set of buildings as one is likely to find anywhere this side of the Atlantic Ocean".

Landscapes have a psychological capacity. Space, loneliness, and uncrowded solitary experiences are attractive landscape features to European visitors (Macpherson Research, 1998). Interestingly these features also contributed to a feeling of personal safety for many single women travellers, a growing market segment. At certain times of the year some of our popular landscapes and features can be seriously overcrowded by any measure. Yet as noted by Fitzpatrick Associates (pers. comm.) about tourism in the Irish landscapes 'to differentiate between the unsustainable and the merely annoying, and to acknowledge that one person's overcrowding may be another person's "craic" is a significant challenge'. Less than 1 per cent of those surveyed on the Glen Shiel Hills said that their enjoyment was seriously affected by the numbers of people met (Herries, 1999). Hillwalkers surveyed on Lochnagar and Upper Deeside stated that the presence of other hill users was only a minor problem (Mather, 1998).

In the process visitors undertake when deciding to visit a destination, landscape is frequently a primary motive. The landscape has to suit the visitor's needs, for example in terms of sight seeing, activities, peace and quiet, or wildlife. Once that choice is made, secondary filters relating to type and availability of transport, accommodation and other facilities begin to influence visitor decisions. Visitors attracted to a place by its landscape setting will on arrival still ask what is there to do, as the landscape becomes a backdrop to other activities. These activities and services then impact on the landscape and may erode the very features that drew the visitors to the area initially. This presents a curious dichotomy. Clearly landscape alone, for all its importance in causing a visitor to choose a destination, does not necessarily mean that the visitor will value that landscape. Visitors need access into the landscape and they need services and facilities while they are there. Herein lies the challenge of conserving the character of an area while encouraging visitors to come. Promotion can contribute to pressure, as for example 'virtually everyone [German visitors] wants to go to Skye' (Macpherson Research, 1998). This highlights the need for tourism and landscape interests to work together, possibly to increase the capacity of the landscape to absorb tourism without eroding its character or conservation status.

How is this capacity to be managed? A number of mechanisms have been tried, but in Scotland only on the Nevis Range has there been a serious attempt to manage capacity in a wildland setting. It is practically difficult and socially unacceptable in many quarters to charge people in order to control capacity in all but the most managed of landscapes. Ownership of the landscape is rarely in the hands of those that gain direct financial benefit from the visitor. However, changing circumstances in agricultural support mechanisms, farmgate prices and access legislation will cause land managers to reconsider options for enhancing revenue through tourism opportunities. As most tourism businesses only indirectly benefit from the landscape there can be indifference to landscape issues, even if it is the visitor's primary motive for visiting the area. The landscape owner or manager may even perceive a threat through damage or loss of privacy from the visitor gazing on their landscape and consuming their resource for free. As a consequence the visitor may experience an unwelcoming attitude, at odds with the traditional hospitality being sought by the tourism industry. In the eyes of some landowners, visitors cause management problems and may be seen as detrimental to other possible land uses.

7.5 The role of the tourism industry in influencing landscape change

Recognition of the strategic value of landscapes and the wider environment is beginning to emerge in tourism planning documents, although the economic linkages are still underplayed. In spite of the 'Unique Selling Point' of Scottish tourism vested in Scotland's landscape character, the tourism industry has been slow to recognise the need to safeguard or understand this key asset better. If sustainable tourism practices are to be pursued, there is a need for the industry to be part of the management processes, influencing use of its key asset. The industry itself should examine visitor perceptions underlying the 80 per cent figure to allow the individual tourism operator or community to realise the potential of the local landscape to help their business. Quotations such as " Our landscapes are of international renown and attract visitors from throughout the world" (Anon., 1998a, p. 52); " ... the [tourism] development strategy must be founded on the market appeal of Scotland's scenery and environment ..." (Scottish Enterprise, pers. comm.); and " The natural environment is Scotland's number one tourism resource, both in terms of its physical scope and the opportunities it represents" (Anon., 1994, p. 31) convey these thoughts.

Market surveys hint at public preferences. The Highland Visitor Survey of 1997 identified a visitor spend per person per night in Sutherland of £62 compared with £35 per person per night in Badenoch and Strathspey. Is landscape a factor in this difference? Sutherland has far fewer opportunities than Badenoch and Strathspey for visitors to spend money and has a lower average length of stay. However, Sutherland has a greater proportion of visitors who spend more, including those from overseas, and those who come from higher social categories. There is no direct economic link between landscape quality and visitor spend, but there is an indication that landscape management does influence certain market segments and may contribute to increased spend.

Tourism depends on the landscape, of that there is no doubt. Tourism has at times been guilty of landscape despoilation, usually for reasons deemed economically and socially sensible at the time. There is now an increasing realisation amongst the tourism industry that it is " economically counterproductive to allow tourism development and activity to spoil the landscape people came to enjoy" (Anon., 1995b). Tourism is often a target for criticism because it is an obvious activity and visitors by definition are not local to an area, creating a 'them and us' mentality. Tourism should be an ally in landscape conservation and the promotion of biodiversity; hence appropriate tourism and leisure should not be squeezed out by unreasonable controls and protectionism.

" This is especially the case when calls for the restriction of recreational activities are based on unfounded ideas of environmental damage or the desire for a tidy and picturesque landscape. It must be remembered that much of Britain's heritage was not sustainable in its originally designed form, anyway. Traditional stone walls in upland farming areas were never sustainable as field boundaries; stone castles, after a very short military life, crumbled to picturesque ruins. These and countless others have lived a much longer life as landscape features and tourist attractions than as functional artefacts" (Anon., 1995b, paragraphs 174 and 177)

There are many other impacts on landscape over and above tourism. The cumulative effect of primary industries, settlement and transport will change and clutter a landscape, and their impact on tourism will be unnoticed until the visitors stop coming or returning. There are pressures for uniformity and sameness related to a poor understanding of our natural and cultural heritage (Anderson, 1998). Equally, landscapes are part of dynamic processes, which occur well within human life spans. Dramatic change can almost be an attraction in itself. Major natural traumas can create new landscape for the visitor to gaze upon. More controlled, man-made intervention, for example, in the flooding of valleys behind dams and large-scale afforestation, have been used to alter the landscape and in some cases have created the opportunity for a major expansion in tourism activity. The ideas behind sustainable tourism should not lead to conserving landscapes in an artificial time freeze frame state.

Area Tourism Strategies are the mechanisms that will be used across the 14 Scottish Area Tourist Boards to provide local direction to the industry. Area Tourism Strategies need to take more account of those outwith the industry who have influence over landscapes, including planning bodies, public agencies and land managers. The Chairman of the Highlands of Scotland Tourist Board said

" The Highlands of Scotland have a number of natural attributes – the beauty of the scenery, the purity of the environment and the hospitality of the people – we must ensure that these are preserved for the future" (Anon., 1997a, p. 2).

Area Tourism Strategies need to consider the use and value of the landscape by particular market segments and to identify the strengths of their particular area e.g. coastlines and peninsulas, moorland and inland plateaux. There is a danger that visitors entranced by the definitive Highland landscape promoted as Scottish, find a different, possibly less satisfactory, product in Ayrshire or the Borders. Gold and Gold (1995) contend that this Highland imagery has subsumed all other imagery of this diverse country and may limit the potential of tourism marketing. Tourism may be the only effective means or justification for financially supporting the maintenance of valued countryside features (Selman, 1996).

7.6 Conclusions

All Area Tourism Strategies should pursue a core strategic objective which promotes and safeguards landscape in its natural and cultural context and recognises local distinctiveness. At a practical level tourism strategies should continue to minimise the effect of tourism on the landscape through

- better partnership with public and private land managers;
- environmental improvement schemes to create more attractive areas for visitors;
- attempting to define visitor capacity management better;
- improved landscape considerations for roads, waterways, parking facilities, footpaths and transport services;
- general encouragement of more sustainable design (energy, water, waste) in the development process of tourism infrastructure and facilities;
- improved landscape interpretation; and
- training and market research related to tourism landscape issues.

Landscape is fundamental to our self image and it has high economic significance. Landscape planning is perhaps too distant from the changing needs of the tourism industry. To conserve our primary tourism asset and move towards a more sustainable approach, there is a need for a closer dialogue between the tourism industry and landscape professionals at both a strategic and local level.

Note: the views expressed in this paper are my own and should not be considered policy of the Tourism and Environment Forum.

References

Anderson, R .1998. Rural housing. *Newsletter of the Association for the Protection of Rural Scotland*, **32**, 6.

Anonymous. 1992. *Tourism and the Scottish Environment. A Sustainable Partnership.* Edinburgh, Scottish Tourism Co-ordination Group.

Anonymous. 1994. *Scottish Tourism Strategic Plan.* Edinburgh, Scottish Tourist Board.

Anonymous. 1995a. *Tourism factsheet.* Edinburgh, Scottish Tourist Board.

Anonymous. 1995b. *The Environmental Impacts of Leisure Activities.* London, HMSO.

Anonymous. 1996. *Agenda 21 for the Travel and Tourism Industry.* London, World Travel and Tourism Council.

Anonymous. 1997a. *Annual Report.* Inverness, Highlands of Scotland Tourist Board.

Anonymous. 1997b. *Highland Visitor Survey.* Inverness, Highlands of Scotland Tourist Board.

Anonymous. 1997c. *Overseas Brochure – When will you go?* Edinburgh, Scottish Tourist Board.

Anonymous. 1997d. *Tranquillity Maps 1960's and 1996: Aberdeen to Inverness.* Edinburgh, Association for the Protection of Rural Scotland.

Anonymous. 1998a. *Towards a Development Strategy for Rural Scotland: the Framework.* Edinburgh, The Scottish Office.

Anonymous. 1998b. *Cost Effective Landscape: Learning from Nature.* Edinburgh, The Scottish Office.

Anonymous. 1998c. *Greenways.* Herald Newspaper, **18 August 1998**, p. 11.

Baxter, C. and Thompson, D.B.A. 1995. *Scotland - Land of Mountains.* Grantown, Colin Baxter Photography.

Farrell, B.H. and MacLellan, R.W. 1987. Tourism and physical environment research. *Annals of Tourism Research*, **14**, 1-16.

Findlay, I. 1976. *The Central Highlands.* London, Batsford.

Gold, J. and Gold, M. 1995. *Imagining Scotland: Tradition, Representation and Promotion of Scottish Tourism since 1750.* Aldershot, Scolar Press.

Gourlay, D. and Slee, W. 1998. Public preference for landscape features: a case study of two Scottish Environmentally Sensitive Areas. *Journal of Rural Studies*, **14**, 248-263.

Herries, J. 1998. *Glen Shiel hill walking survey 1996: 3 August-3 November 1996 - report of findings.* Scottish Natural Heritage Research, Survey and Monitoring Report No. 106.

Hunter, J. 1995. *On the Other Side of Sorrow.* Edinburgh, Mainstream.

Innes, N. 1998. Tourism Transport in Scotland. In MacLellan, R. and Smith, R. (Eds.) *Tourism in Scotland.* London, International Thomson Business Press. pp. 170-186.

Leslie, A. 1997. *The Which? Guide to Scotland.* London. Which? Ltd.

MacLellan, R. 1998. Tourism and the Scottish Environment. in MacLellan, R. and Smith, R. (Eds) *Tourism in Scotland.* London, International Thomson Business Press. pp. 112-134.

Macpherson Research (1998). *Perceptions and experiences of access to the Scottish countryside for open air recreation of visitors from mainland Europe.* Scottish Natural Heritage Research, Survey & Monitoring Report No. 32.

Mather, A. 1998. *East Grampians and Lochnagar Visitor Survey 1995: overview.* Scottish Natural Heritage Research, Survey & Monitoring Report No. 104.

Middleton, V.T.C. and Hawkins, R. 1998. *Sustainable Tourism a Marketing Perspective,* Oxford, Butterworth Heinemann.

Murray, W.H. 1962. *Highland Landscape.* Aberdeen, Aberdeen University Press.

Selman, P. 1996. *Local Sustainability: Managing and Planning Ecologically Sound Places.* London, Paul Chapman.

Smallman, T. and Cornwallis, G. 1999. *Scotland.* Victoria, Australia, Lonely Planet.

Urry, J. 1990. *The Tourist Gaze.* London, Sage.

8 ASSESSING PUBLIC PERCEPTION OF LANDSCAPE IN WALES: A *LANDMAP* APPROACH

J.M. Bullen

Summary
1.   A methodology was developed for assessing public perception of landscape using both qualitative and quantitative information.
2.   The methodology adopted achieved a clear understanding of public perception that could be successfully integrated into the landscape assessment process.

8.1 Introduction

The *LANDMAP* (Countryside Council for Wales, 1996) approach to landscape resource assessment was introduced by the Countryside Council for Wales and the Welsh Landscape Partnership Group. Consultation and pilot testing in three areas within Wales has led to the compilation of the first handbook (Countryside Council for Wales, 1998). *LANDMAP* aims to bring together the process of landscape assessment with a flexibility that allows management decisions to be made at the level of the individual planning application or the much larger level of a broad overview.

The assessment identifies the characteristics of individual landscapes, which are represented as discrete geographical areas, known as 'Combined Aspect Areas'. These Combined Aspect Areas are based on the following criteria which are collectively grouped into three areas.   The *evaluated* aspects include geology/geomorphology/hydrology, visual/sensory/spiritual, vegetation/habitats, historic information and cultural/artistic associations and folklore. The *base* aspects of settlement/development and rural land use and the *informative* aspects comprising of public perception and information obtained from designation/strategy documents.

8.2 Public Perception

The public perception aspect aims to involve the public in the landscape assessment process by seeking to understand their perceptions and preferences of landscape elements and types. In particular, the study focuses on understanding the public's sense of landscape identity, identifying which features and landscapes are of value to the local community, their perceptions of landscape value and of landscape types in the study area and finally which landscapes the public wish to conserve, enhance or change.

Whilst the study sets out to address these issues it also responds to the factors that have been identified as influential in the individual's perception of the landscape.   The combination of the responses to questions and the general background questions completed by each participant in the study allow for the following additional factors to be analysed:

biological origins (place of birth or where one grew up), sense of place, cultural associations, age, familiarity and feelings evoked.

8.3 Methodology

The methodology developed uses two different approaches, a household questionnaire and structured focus group discussions.

A carefully structured questionnaire was developed to meet the aims of the study and the requirements of a particular study area. Household respondents were targeted to complete the questionnaire in an interview situation with a trained surveyor. A preliminary letter was sent to respondents outlining the study and their role within it.

Respondents were selected in a manner that ensured a representative sample of the population whilst also addressing the practical constraints of time and financial resources. Five electoral wards were chosen in order to develop clustered samples. These wards were selected at random from previous classification surveys of the wards to allow different categories to be compiled. From within these wards individual respondents were then selected at random. Retrospective stratification was employed in the Vale of Glamorgan and Cardiff to balance the sample gained with local population characteristics (e.g. age, sex, ethnic group, Welsh speaking). This procedure is recommended where possible.

The questionnaire used prompt cards and landscape photographs as visual stimuli. In addition to setting questions to address the issues outlined in Section 8.2, several more general, background questions were also asked.

Focus group discussions were conducted with the following nine groupings: children, youth, men, women, senior citizens, people with special needs, unemployed, rural land use and tourists. Each focus group included 8 to 12 participants, a facilitator and a secretary. A structured one and a half hour discussion covered the aims of the study, and also provided an opportunity to explore and develop views and preferences. The same photographs were displayed, but as slides. In addition, participants completed a brief questionnaire of general background questions in order to develop a profile of the participants.

The methodology was pilot tested in Llŷn, North Wales. Comprehensive studies have been undertaken for the Vale of Glamorgan, Gwynedd and Cardiff.

8.4 Output examples

Two examples are presented from the Vale of Glamorgan Report (Bullen *et al.*, 1998) to demonstrate the type of information that is derived from the studies. The examples are extracts from selected questions and reflect the complimentary nature of the two approaches in generating both quantitative and qualitative information.

8.4.1 Study example 1

Respondents completed an assessment of up to seven photographs of the landscape types represented in the study area. By analysing the specific comments made by the public about each photograph, an understanding of the importance of the landscape type and its associated landscape elements can be gained. The aspect specialists can use this information to draw together public perception and professional opinion for specific landscape types.

For example, the responses associated with photograph 1 (Plate 13 and Table 8.1) demonstrate that the landscape is positively perceived, particularly with respect to the

Table 8.1. Comments in response to the question "Why do you like/dislike Photograph 1?".

Questionnaire Responses		Examples of Focus Group Responses
Traditional village/farmland scene	56%	Green and pleasant
Views	22%	Countrified
Unspoilt	11%	Tranquil and idyllic
		Safe
		Bit boring

'traditional' and 'unspoilt' character. More detailed responses highlight the importance of villages and farmland as a distinctive feature of the study area. The responses also indicate the aesthetic and esoteric qualities associated with a particular landscape type.

In photograph 2 (Plate 14 and Table 8.2) respondents were most influenced by the colours in the landscape, but also positively assessed the farmland elements (fields, trees and woodlands) and the 'unspoilt' qualities of the landscape.

Table 8.2. Comments in response to the question "Why do you like/dislike Photograph 2?".

Questionnaire Responses		Examples of Focus Group Responses
Colours	36%	All aspects of the countryside
Farmland landscape	26%	Forest, nice for walks
Views	25%	Contrasting colours in the fields
Unspoilt	23%	Freedom, where you could wander
Naturalness	15%	Dislike cut trees on horizon
		Wildness at front
		The flowers and the trees

8.4.2 Study example 2

Responses to the questions "what needs conserving?" and "what are the greatest threats?" in the study area indicate public concern for their local landscapes. This can usefully be included in planning and management decisions. The range of responses from the questionnaire indicates the strength of feeling and the issues that are important to the public. Table 8.3 lists some of the responses obtained.

(a)

(b)

(c)

Plate 1. Three images of Scotland's landscape, indicating a designed landscape, recreation in a lochside landscape, and a landscape portraying seasonality in the hills (see chapters 1). (a) Cortachy Castle (Photo: P. & A. Macdonald), (b) Loch Lomond (Photo: L. Gill, SNH), Glen Lyon (Photo: L. Gill, SNH).

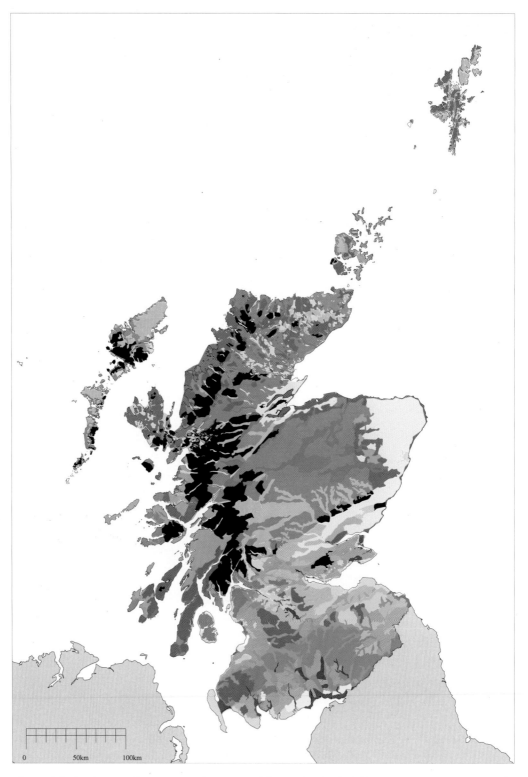

Plate 2. The level 3 map demonstrating the groups of the Scotland-wide landscape character assessment (see Chapter 1; the key is on the opposite page).

Key: Scotland Schedule Level 3

HIGHLANDS AND ISLANDS

1. Farmlands and Estates of the Highlands and Islands
2. Flat or Rolling, Smooth or Sweeping, Extensive, High Moorlands of the Highlands and Islands
3. High Massive Mountain Plateau of the Cairngorms
4. High, Massive, Rolling, Rounded Mountains of the Highlands and Islands
5. High, Massive, Rugged, Steep-Sided Mountains of the Highlands and Islands
6. Highland Cnocan
7. Highland Foothills
8. Highland Straths
9. Highland and Island Crofting Landscapes
10. Highland and Island Glens
11. Highland and Island Rocky Coastal Landscapes
12. Highland and Island Towns and Settlements
13. Knock or Rock and Lochan of the Islands
14. Low Coastal Hills of the Highlands and Islands
15. Low, Flat, and / or Sandy Coastal Landscapes of the Highlands and Islands
16. Moorland Transitional Landscapes of the Highlands and Islands
17. Peatland Landscapes of the Highlands and Islands
18. Rocky Moorlands of the Highlands and Islands
19. Rugged, Craggy Upland Hills and Moorlands of the Highlands, including the Trossachs
20. Sea Lochs of the Highlands and Islands

LOWLANDS

21. Agricultural Lowlands of the North East
22. Coastal Hills Headlands Plateaux and Moorlands
23. Coastal Margins
24. Coastal Raised Beaches and Terraces
25. Drumlin Lowlands
26. Flatter Wider Valleys and Floodplains of the Lowlands
27. Low Coastal Farmlands
28. Lowland Coastal Flats Sands and Dunes
29. Lowland Coastal Landscapes of the North East
30. Lowland Hill Margins and Fringes
31. Lowland Hills
32. Lowland Loch Basins
33. Lowland Plateaux and Plains
34. Lowland River Valleys
35. Lowland Rolling or Undulating Farmlands, Hills and Valleys
36. Lowland Urbanised Landscapes
37. Lowland Valley Fringes
38. Narrow Valleys in the Lowlands
39. Rocky Coasts Cliffs and Braes of the Lowlands
40. Rocky Volcanic Islands

UPLANDS

41. Foothills and Pronounced Hills
42. High Plateau Moorlands
43. Rugged Granite Uplands
44. Rugged Moorland Hills
45. Smooth Upland Moorland Hills
46. Upland Basin
47. Upland Fringe Moorland
48. Upland Fringe Valleys and Farmlands
49. Upland Glens, Valleys and Dales
50. Upland Hills, The Lammemuir, Pentland and Moorfoot Hills
51. Upland Hills, The Southern Uplands and Cheviots
52. Upland Igneous and Volcanic Hills The Ochil, Sidlaw, Cleish and Lomond Hills

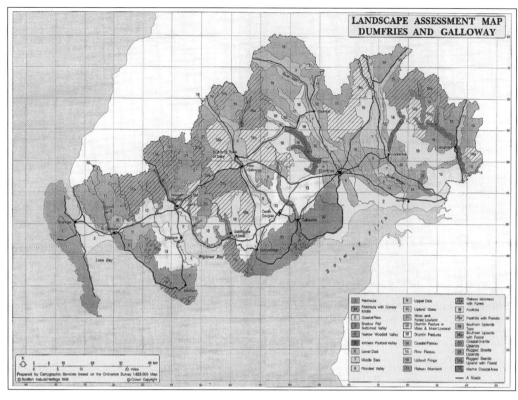

Plate 3. Dumfries and Galloway landscape assessment: the landscape character types (see Chapter 2).

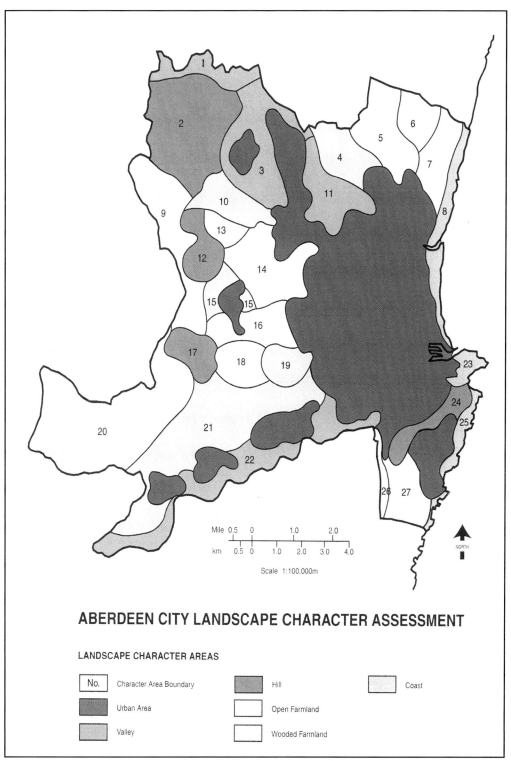

Plate 4. The landscape character assessment of the City of Aberdeen. The landscape character areas and types are explained in the text (see Chapter 2).

Plate 5. Examples of four photographs used in the questionnaire, showing (a) montane, (b) coastal, (c) man-made artefacts and (d) native woodland (see Chapter 4). (Photo: D. Habron).

Plate 6. Part of the Alnarp landscape laboratory; the aerial photograph, taken in 1994, shows an overview of the one kilometre long, open water-course, and its three different small water bodies. These are equal in size but with three very different designs. Beside the water-course there are some newly established meadow corridors and a newly planted multi-functional plantation (see Chapter 5). (Photo: A. Bramme).

Plate 7. Oxhagen is one of the few well-preserved grazed area in Scania, and is one of the main reference landscapes and long term research areas. Throughout history the pastoral landscape has been one of the most popular landscape types, but as such it has been viewed as very static through time. The aerial photograph, taken in 1996, gives an overview (see Chapter 5). (Photo: A. Bramme).

Plate 8. A view of Oxhagen in 1983 (see Chapter 5). (Photo: R. Gustavsson).

Plate 9. A similar view of Oxhagen in 1998 to that in Plate 8. After an interval of fifteen years, this photograph illustrates the recent simplification processes (see Chapter 5). (Photo: R. Gustavsson).

Plate 10. Tjärö, an island in the south-east part of Sweden, is characterized by its landforms, rocks, grazed grasslands, and tree- and shrub-rich landscapes. It has a cultural identity with traditional buildings, fences, stone walls and pollarded trees. Tjärö is used every year for educational programmes related to landscape design and land management (see Chapter 5). (Photo: A. Bramme).

Upland Glens (10)

Existing Character

Oak, ash and birch woods on valley floor

Remnant semi-natural woodlands on steep gullies

Single trees around dwellings

Screes and incised channels

Distinctive sculpted landform

Flat valley bottom with improved pasture

Isolated, rectangular shaped conifer woodlands on valley sides unrelated to landform

Landscape Character

The *Upland Glen* landscape is found along the upper reaches of rivers which cut through the upland areas to the north and east of Dumfries and Galloway. The glens have the pronounced 'U'-shape of glaciated upland valleys and the height of the glen sides (rising to over 300m) creates strong enclosure. There are long and often awe-inspiring views funnelled along the glen, uninterrupted by irregularities in landform or tree cover.

Open rough pasture predominates, with a gradual transition to heather moorland and screes on the upper slopes. There are some improved pastures on the valley floor, with occasional mature oak woodlands on the lower slopes. Forestry has been introduced into parts of the upland glen landscape, usually covering the side slopes and leaving the floor of the glen clear. Dry stone dykes are an important characteristic of the lower slopes. Development is sparse, consisting of isolated farmsteads and occasional houses and the landscape retains a special 'wildland' character.

Local Ecology

The extensive open rough pastures of the upland slopes may be important acidic grassland habitats, with rare plant communities, and the heather moorland is also extremely valuable in ecological terms. On steeper slopes there are extensive screes of botanical, invertebrate and ornithological interest. In addition, the scattered mature oak woods, which are found particularly on the lower slopes of the glens, provide a habitat for a wide variety of flora and fauna; many are remnants of the once extensive semi-natural and ancient woodlands. The fast-flowing streams are often lined with riparian woodland and associated wet grasslands, of botanical and ornithological importance.

Plate 1. Sheet 10 for Upland Glens landscape character type from the Landscape Design Guidance for Forests and Woodlands in Dumfries and Galloway. The plate shows the location plan of landscape character type areas and an analysis of the existing landscape character, illustrated by a typical sketch and photographs (see Chapter 6).

Upland Glens (10)

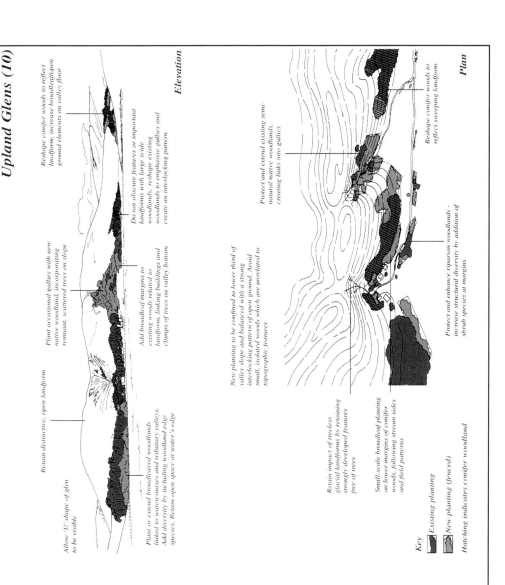

Reshape conifer woods to reflect landform, increase broadleaf/open ground elements on valley floor

Retain distinctive, open landform

Plant occasional gullies with new native woodland, incorporating remnant, scattered trees on slope

Do not obscure features or important landforms with large scale woodlands; reshape existing woodlands to emphasise gullies and create an interlocking pattern

Add broadleaf margins to existing woods related to landform, linking buildings and clumps of trees on valley bottom

Allow 'U' shape of glen to be visible

Plant or extend broadleaved woodlands linked to watercourses and tributary valleys. Add diversity by including woodland edge species. Retain open space at water's edge

Elevation

Protect and extend existing semi-natural native woodlands, creating links into gullies

New planting to be confined to lower third of valley slope and balanced with a strong interlocking pattern of open ground. Avoid small, isolated woods which are unrelated to topographic features

Reshape conifer woods to reflect sweeping landform

Protect and enhance riparian woodlands - increase structural diversity by addition of shrub species at margins

Retain impact of treeless glacial landforms by retaining strongly developed features free of trees

Small-scale broadleaf planting on lower margins of conifer woods, following stream sides and field patterns

Plan

Key

Existing planting

New planting (fenced)

Hatching indicates conifer woodland

Guidance Proposals

Forestry Opportunities and Constraints

- This upland landscape has an open, wild character, with sculptural glacial landforms. Such areas of dramatic landform are visually sensitive and there is likely to be some limited scope for new woodlands.

- There are scattered examples of conifer woodlands which are poorly sited and designed, detracting from the characteristic scale and qualities of the scenery. The lack of topographic variation on the side slopes makes the integration of new woodlands particularly difficult. Planting is more suited to those upland glen areas which lack dramatic glacial profiles and which are less valued for their scenic qualities. Larger scale planting on upper slopes should only be considered where glacial features are less significant and where the semi-natural vegetation is least valued.

Design Guidance

- On valley sides, woodlands should be flowing, asymmetrical shapes, seeking to highlight hollows and watercourses. Major ridges and spurs should remain unplanted to emphasise their dramatic topography. On the valley floor, woodland shapes can follow local field patterns.

- Small to medium-scale planting is appropriate, predominantly at lower levels where it should extend existing woodland patterns.

- Where planting is acceptable on upper slopes, its overall shape and species patterns should relate to the landform. On lower slopes varied patterns of native and conifer woodlands should be designed to emphasise the topography. Managed open ground will be an important element in this zone to avoid obscuring distinctive topographic features and to retain open views.

- Planting should not extend to more than one third of the landscape visible from principal viewpoints and the valley floor and should be concentrated on the lower slopes. A strong, interlocking pattern of woodland and open ground is important.

- Conserve the distinct, sculptural landforms and retain long, scenic views to the open upland landscape.

Cross References:

Southern Uplands (19) and *Foothills (18)* - adjacent in upland areas; *Middle Dale (7), Upper Dale (9)* and occasionally, *Intimate Pastoral Valleys (5)* - adjacent in lowland areas.

Plate 12. As Plate 11, but the reverse side of Sheet 10 for Upland Glens landscape character type. The plate provides succinct design guidance describing how new woodland and forests can be integrated within the existing landscape character, illustrated both by a sketch elevation and plan view (see Chapter 6).

Plate 13. Photograph 1 - the rural village used in the first study example (see Chapter 8). (Photo: Jill Bullen).

Plate 14. Photograph 2 - the wooded vale landscape also used in the first study example (see Chapter 8).
(Photo: Jill Bullen).

Plate 15. Original picture: characteristic view of the village of Krokshult, with red cottages, wooden fences, mounds of stones and a mosaic of small fields and pastures (see Chapter 9). (Photo: C. Hägerhäll).

Plate 16. Manipulated picture: the red cottages and wooden fences have been erased using Photo Shop 3.0 (see Chapter 9). (Photo: C. Hägerhäll).

Plate 17. An example of a new countryside house development that is poorly located in its landscape, as explained in the text (see Chapter 10). (Photo: J. Moir, D. Rice, and A. Watt).

Plate 18. An example of a new countryside house development that is well located in its landscape, as explained in the text (see Chapter 10). (Photo: J. Moir, D. Rice and A. Watt).

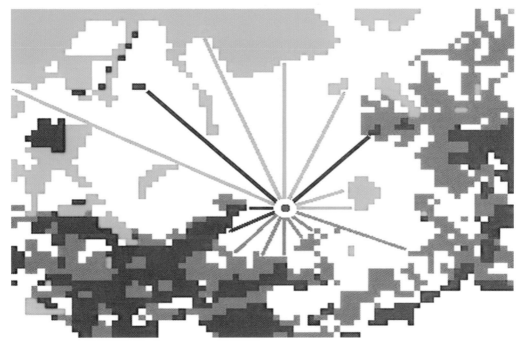

Plate 19. Illustration of the radial method. Different metrics can be derived from parameters such as radial length, nature of obstacles, sequence of lines, radial length variation, land cover types crossed by the lines, etc. The illustration is an extract of the land cover dataset of Flanders. Each pixel is 20m square. Basic land cover classes are woodland (green), densely built (dark red), sparsely built (pink) and bare soil and low vegetation (white) (see Chapter 13).

Plate 20 Visual complexity as measured by a regular sample of 64 view lines in each pixel of open space. Complexity increases with lighter shades of blue. Green patches are woodland, red patches are built up, and white lines are major roads (see Chapter 13).

(c)

(d)

Plate 21. Examples of 3D landscape visualizations that can be created today with a variety of geographic data and real landscape objects. (a) 3D rendering of 10 m resolution satellite imagery over a digital terrain model using World Construction Set (© Questar Productions). (b) A 3D visualization using ERDAS VGIS with 50 cm resolution aerial photography on a 50 m DTM (© The GeoInformation Group). (c) Visualization using satellite imagery but incorporating objects, here showing the flightline of an aircraft through the model (image reproduced with permission by ERDAS UK). (d) A rendered image of the globe using low resolution (4 km) satellite data merged with a global DTM and wrapped around a sphere (© GlobalVisions images) (see Chapter 14).

Plate 22. A frame from an animated drive along the Great Ocean Road, Victoria, Australia. A terrain model and the introduced objects are rendered from an identical view-point as the video frame. Terrain was rendered in black to hide the buildings, as appropriate, but allowing the video to project through (see Chapter 16).

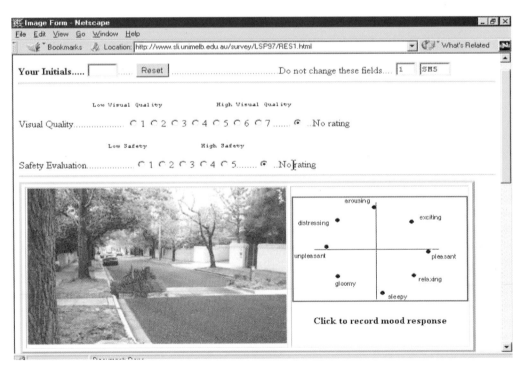

Plate 23. A sample web page incorporating an interactive perception survey. Users record their assessment of visual quality and safety. When they click a location on the affect response diagram (Russell and Lanius, 1984) the survey checks that their responses are complete and valid and then moves on to the next image (see Chapter 16).

Plate 24. Aerial photograph of Parks of Aldie, Kinross, from the east, showing Aldie Castle and its policies at the bottom left, and the Improvement Period fields and plantations covering the rest of the photograph within which there are traces of relict pre-improvement settlement and rig cultivation (see Chapter 17). (Photo: RCAHMS Crown Copyright ©).

Plate 25. Aerial photograph of Ettleton Churchyard and Kirkhill, Liddlesdale, Borders, from the east showing extensive pre-improvement relict landscapes. At the bottom of the photograph there are the rectangular fields of the Improvement Period allotments of Newcastleton, above which there are the curvilinear banks of the pre-improvement fields which run up to and over-ride a curving medieval intake dyke. The hilltop is marked by the ramparts of a prehistoric fort (see Chapter 17). (Photo: RCAHMS Crown Copyright ©).

Table 8.3. Comments in response to the questions on 'landscapes to conserve' and 'threats to the landscape'.

"What needs conserving?"		"What are the greatest threats?"	
Woodlands	33%	More development	39%
Fields	22%	Housing	29%
Greenfield sites	21%	Industry	19%
Coast	15%	Roads/traffic	16%
Rural landscape	12%		

8.5 Application of results

By analysing the responses from both the questionnaire survey and the focus groups it becomes evident which landscapes are valued over others and the types of responses certain elements in the landscape stimulate. The examples highlight the public's appreciation for landscapes associated with managed farmland, with villages and woodlands being integral and important features of these landscapes. Study example 2 demonstrates the public's desire to conserve these woodlands, fields and rural land and their underlying concerns that these landscape features are under threat from development pressure. Not only is housing pressure at the edges of villages and within the rural landscape seen as a threat, but the size and scale of new developments, which are out of character with the local landscape, are also seen as threats. Very often it is the further probing of information that the focus group sessions allow that can not only enhance the understanding of public perception but can also highlight any divergence in opinion between the different sectors of the public. Information of this nature can be constructively included to complement the expert analysis of the landscape and to ensure that the public's opinions can be incorporated into practical action plans and guidelines which take into account the responses from all sectors of the community. In this instance, guidelines which promote the conservation and enhancement of landscapes with these features would be appropriate, as would recommending guidelines for developments to reflect local identity and style.

8.6 Conclusion

The *LANDMAP* methodology is successful in achieving results that develop a clear understanding of landscape perception, with opinions being drawn from all sectors of the local community. The two approaches used to elicit public opinion are comparable and complimentary in nature and take full advantage of both quantitative and qualitative information whilst meeting the practical problems of sample size.

The results can be included in the early stages of the assessment process by feeding into the aspect specialists' reports. However, they can also be successfully considered at the later stages, once the Combined Aspect Areas have been defined, by using these areas as the basis for the photographic assessments.

Acknowledgements

The research work was developed and completed jointly with Alister Scott and Euryn Jones, University of Wales, Aberystwyth. My thanks are extended to the Vale of Glamorgan Council for their permission to reproduce the photographs and examples used.

References

Bullen, J.M., Scott, A.J. and Jones, E.M. (1998). Public perception of landscape in the Vale of Glamorgan. Unpublished report to the Vale of Glamorgan Council.

Countryside Council for Wales (1996). *The Welsh landscape: our inheritance and its future protection and enhancement: CCC16.* Bangor, Countryside Council for Wales.

Countryside Council for Wales (1998). LANDMAP: a multi-purpose approach to supporting landscape decisions. Unpublished manuscript.

9 ASSOCIATIONS TRIGGERED BY SPECIFIC LANDSCAPE CHARACTERISTICS

Caroline M. Hägerhäll

Summary

1.   This study examined urban people's mental image of the Swedish traditional cultural landscape. There are three particular results.
2.   The 'Swedishness', as well as the idyllic and cultural heritage aspects of this landscape, are embedded in the characteristic red cottages and wooden fences.
3.   The cottages and fences increase the attractiveness of this landscape as a holiday destination.
4.   The cultural landscape is associated with summer. The high frequency of responses mentioning summer is unlikely to be an effect only of the weather in the visual stimuli. Cultural factors, such as the association with holidays, are probably involved.

9.1 Introduction

Man-made landscape features, in general, have been found to attract special interest and affect human preference for a landscape (Kaplan and Kaplan, 1989). Red cottages and wooden fences are not only man-made, but characteristic features of the traditional cultural landscape in Sweden. Furthermore, these objects have been, and are still, frequently used in marketing publicity and propaganda, which is likely to influence people's views both of the objects themselves and the landscape in which they are situated. The idea of the red cottage as the good and genuine Swedish home can be traced back to the political rhetoric at the turn of the century (Edling, 1996).

The story of the red paint, so commonly used that the red cottage has become a symbol of Sweden, started in the early 17th century, when the first red paint was produced. Sweden had little silver but plenty of copper, which was used, for instance, in coins. The red pigment was a by-product from copper mining in the large mine in Falun, which has also given the paint its name, *Falu rödfärg*. Inspired by the red brick houses on the European continent, the noblemen and soldiers returning from the Thirty Years' War, in the mid 17th century, started to paint their houses in red paint. For the common man it was still much too expensive to paint the house. The shortage of timber in the 18th century promoted the use of the red paint, since it protected the wood from rot and thus made the wooden houses last longer. Many important government buildings, churches and rectories were protected with the red paint. The fashion of the upper classes thus slowly spread to the cities and during the 19th century to the farmers in the countryside, who had started to become richer. During the national romantic period in the early 20th century, and the own-your-own-home movement in the 1930s, the use of the red paint boomed and 2,000 tonnes of paint pigment was produced annually in Falun. Sweden had become the country of the red cottages.

This study examines the importance of the red cottage and the wooden fences to the mental image of the cultural landscape by comparing people's associations to two different versions of the same picture.

9.2 Method

9.2.1 Visual stimuli

A colour slide, depicting a characteristic view of the village of Krokshult in the region of Småland (location shown in Figure 9.1) with red cottages, fences, mounds of stones and a mosaic of small fields and pastures, was scanned into the computer. The original picture (Plate 15) was used as the first visual stimulus. For the second visual stimulus (Plate 16), the houses and wooden fences were erased from the original picture using Photoshop 3.0. The erased areas were filled in with the adjacent vegetation. In all other aspects the pictures were identical. The two pictures were then printed out on glossy photo quality printer paper and used in the survey.

9.2.2 Procedure

People visiting a large shopping centre outside the city of Malmö in southern Sweden were shown one of the two pictures and asked what the picture made them think of. For Plate 15 a total of 98 people was interviewed, 47 males and 51 females, with ages ranging from 7 to 72 years. For Plate 16 a total of 85 people was interviewed, 33 males and 52 females, with ages ranging from 8 to 76 years. The keywords in the subjects' answers were written down by the interviewers and later analysed with a self-organising neural network for text analysis, CATPAC II Plus for Windows (Woelfel and Stoyanoff, 1995).

So far the software has been used mostly in marketing research and hence the body of published empirical studies is limited. The program is designed to read ASCII text and learn the interrelationship between the words in a section of text. However, the study reported here only made use of the simple word counts and frequency tables that the program also provides. Tests of the more advanced analyses of interrelationships between words, as well as a review of the use of the program and its performance, can be found in Hägerhäll and Schifferl (1999).

9.3 Results

Listed in Table 9.1 are the 25 most frequently mentioned words for the two pictures. The three most frequently mentioned words for both pictures were *Småland*, *Summer* and *Countryside*. This shows that many people recognised the region, which should be noted is a region close to where the subjects lived. The responses *Summer* and *Countryside* could, to some extent, reflect an effect of just reporting what can be readily seen in the picture. However, it is unlikely that an equally high frequency of *Summer* would be obtained for just any picture taken during the summer season. This indicates that cultural factors connect this landscape especially to summer.

Differences between the pictures were also found. Plate 15 had a larger frequency of words relating to 'Swedishness', history and cultural and national heritage. Good examples

Figure 9.1. Map of Sweden, showing the location of Småland and the village of Krokshult.

Table 9.1. Lists of the 25 most frequent words for the two pictures (Plates 15 and 16) and the percentage of cases in which they were used.

Plate 15		Plate 16	
Original with red cottages and wooden fences		*Red cottages and wooden fences erased*	
Word	*Case percentage*	*Word*	*Case percentage*
Småland (district)	30	Småland (district)	19
Summer	18	Summer	16
Countryside	13	Countryside	11
Cottage	10	Forest	9
Swedish	9	Nature	8
Astrid (author)	7	Landscape	7
Fences	7	Heat	7
Quiet	7	Animals	6
Holiday	7	Nice	6
Summer-cottage	7	Cows	6
Old	6	Skåne (district)	6
Idyll	6	Stone	6
Red	6	Swedish	6
Skåne (district)	6	Drought	6
Childhood	5	Open	6
Grandmother	4	Childhood	5
Native-place	4	Native-place	5
Nature	4	Autumn	5
Peace	4	Quiet	5
Nice	3	Stonewalls	5
Freedom	3	Beautiful	5
Horses	3	Barren	4
Landscape	3	Old	4
Forest	3	Picnic	4
Beautiful	3	Peace	4

are the words *Swedish, Old, Idyll* and references made to the author Astrid Lindgren, who used this landscape as a setting for many of her famous children's books. Plate 16 triggered more associations about nature and animals. Notable is the higher percentage of cases mentioning *Forest* for Plate 16 and the words *Stone, Drought, Heat* and *Barren,* referring to hard soil and climatic conditions.

A most interesting finding is that the picture with houses and fences, much more than the picture only with vegetation, directed people's thoughts to holidays and relaxation. The words *Holiday* and *Summer cottage* were frequent for Plate 15 but were not mentioned at all among the 25 most frequent words for Plate 16. *Peace* and *Quiet* were also mentioned more often for Plate 15 than for Plate 16.

9.4 Discussion

The results indicate that the red cottages and the wooden fences could be important features in the marketing of this landscape as a holiday destination. Further evidence for this was found in the answers to an additional question, asking if people would like to spend their holidays in this landscape. The percentage of 'yes' answers was slightly higher for Plate 15 than Plate 16; with responses of 75 and 67 percent respectively.

It is also clear that, for the urban public, the red cottages and wooden fences are important in the identification of cultural landscapes. The characteristic vegetation, stone mounds and small scale structure, which are products of traditional farming, are not equally strong as indicators of cultural heritage to this group of people. It would be of interest to investigate to what extent there is a difference, in perception and opinions, between farmers and environmental professionals on the one hand, and the urban public, with a decreasing personal experience of the living countryside, on the other.

Acknowledgements

This study was made possible with the financial support of the Swedish Tourist Authority.

References

Edling, N. (1996). *Det Fosterländska Hemmet. Egnahemspolitik, Småbruk och Hemideologi Kring Sekelskiftet 1900.* Stockholm, Carlsson bokförlag.

Hägerhäll, C.M. and Schifferl, E. (1999). Application of neural network algorithms for qualitative empirical research. In Hägerhäll, C.M. *The Experience of Pastoral Landscapes.* Acta Universitatis Agriculturae Sueciae, subseries Agraria, 182, Swedish University of Agricultural Sciences, Alnarp.

Kaplan, R. and Kaplan, S. (1989). *The Experience of Nature.* New York, Cambridge University Press.

Woelfel, J. and Stoyanoff, N.J. (1995). CATPAC: a neural network for qualitative analysis of text. In Woelfel, J. and Stoyanoff, N.J. (Eds.) *Artificial Neural Networks for Advertising and Marketing Research.* Amherst, RAH Press, pp. 51-70.

10   LANDSCAPE CHARACTER: ITS ROLE IN PLANNING FOR NEW HOUSING IN SCOTLAND'S COUNTRYSIDE

John Moir, David Rice and Allan Watt

Summary

1.   Scotland has landscapes of international renown which attract visitors from around the world but which are also important to local communities for economic and social benefits.
2.   If the visual character of the countryside is to be maintained, increasing care is required in the siting and design of new houses.
3.   The availability of landscape character assessments represents an important new tool for planning, designing and guiding new housing in the countryside.
4.   At the moment, application of landscape character assessments in local plan policies in Scotland is limited.
5.   There appears to be a need for guidance on the integrated use of landscape assessments and local building design studies to achieve housing that will 'fit' into the countryside.

10.1   Introduction

The issue of new housing in the countryside has long been a matter of interest to town and country planners. Up to the 1980s the operation of the 1947 Town and Country Planning Act was seen as the fundamental means of protecting agricultural land and preserving open countryside for those who lived there and for the enjoyment of the visitor from the town. Local planning authorities were encouraged to minimise visual impacts by channelling new housing into existing settlements and applying strict controls on developments in the open countryside (Cherry, 1975; Department of Health for Scotland, 1960). In more recent years the protection of farm land has become less important on poorer quality land and farmers have been encouraged to diversify their activities (Scottish Development Department, 1987). At the same time, there have been increasing and sometimes conflicting new demands on the countryside, for example recreation and tourism activities and mineral extraction.

The countryside has continued to be an attraction for Scotland's urban population many of whom, with increasing prosperity and mobility, have sought to live in the countryside (Scottish Office, 1998a). It is difficult to quantify the extent of this demand because the Scottish Office does not have data on the national total of submitted planning applications for single houses in the countryside. Nevertheless, the matter has been regarded as sufficiently important to require the formulation of national planning policy and the recommendation of best practice to planning authorities (Department of Health for Scotland, 1960; Scottish Development Department, 1985; Scottish Office, 1991, 1996). Current government policy is to promote housing initiatives that support the rural economy, embody the principles of sustainable development and enhance the rural environment (Scottish Office, 1998b). Accordingly, the challenge for planning today is to allow development while, at the same time, protecting the visual amenity of the countryside and its natural and cultural heritage.

One means by which this challenge can be effectively met is through the use of landscape character assessments. Their application has increased over the last 10 years in order to improve the quality of decision making and to ensure that judgements take full account of the natural and cultural factors that make one place different from another. The national programme of landscape character assessment of Scotland promotes greater understanding of the varied landscapes throughout the country and provides guidance for managing landscape change.

The purpose of this chapter is to examine the emerging experience of some rural planning authorities in Scotland in the use of landscape character assessment for the control of new housing in the countryside. It focuses upon the development of planning policy intended to safeguard landscape character and reviews planning practice intended to identify visually acceptable sites for new houses. The chapter is structured into the following five sections,

- a review of relevant planning policy development over the last 40 years;
- an explanation of government advice to local planning authorities published in PAN36: *Siting and Design of New Housing in the Countryside* (Scottish Office, 1991);
- an examination of planning authority responses to this government advice;
- an assessment of the recent application of landscape character assessments by planning authorities in development plan policy preparation and development control work; and
- a conclusion containing suggestions for further action.

10.2 Planning Policy Context

Before examining the current application of landscape character assessment techniques in respect of decisions about new housing in the countryside, it is useful to remember how planning policy and practice has evolved over the last 40 years. The Department of Health for Scotland Circular No 40/1960 *New Houses in the Country*, set out three main principles regarding new houses in the country. These were

- in a village, permission for a well-sited and well-designed house will be granted unless the local planning authority has a definite reason to the contrary, e.g. where the site is reserved for an essential local purpose;
- in the open country, new houses will not normally be permitted unless the Development Plan provides accordingly, or unless on the merits of the particular case there is a special need; and
- additions to existing ribbon development or scattered building will not usually be allowed.

Formulation of this policy reflected a growing concern, at the time, that if all the demand for new houses in isolated locations away from villages was allowed, houses would be dotted over all the countryside or strung out along the country roads. Even in the remote areas where demand for new houses was expected to be less, there was a concern that a badly-sited or poorly-designed house might ruin the appreciation of fine scenery for many people. Accordingly, the building of new houses in the open country was to be carefully controlled everywhere, and in some places it was not permitted at all except for people such as farmers and workers who had a special operational need to live in the countryside.

In succeeding years, Development Control Officers in planning authorities applied a series of visual considerations when evaluating the merits of individual proposals which could be approved within the terms of the policy. For example, account was taken of the sphere of visual influence or prominence of a proposal on a given site and the extent to which this might be ameliorated by tree planting or local adjustments of siting to take advantage of visual containment derived from landform and vegetation. A site visible from a number of public viewpoints was less likely to receive planning consent (Angus District Council, 1980; Highland Regional Council, 1977). Neither was it likely that consent would be given for a house which would intrude upon views of a loch from a public road or break a skyline. Whereas decisions on individual proposals took regard of site characteristics, there is little or no evidence to suggest that it was normal practice at that time to take account of the wider relationship between a proposal and the landscape character of the surrounding countryside and its settlement pattern.

In 1985 Government policy for housing in the countryside was reviewed because of three key factors which had emerged since 1960; it was considered that

- the countryside had become more accessible for residence, employment and leisure;
- the types of development required or proposed had become more varied than the new houses with which Circular 40/1960 was mainly concerned; and
- new non-traditional activities were making an important contribution to the rural economy, and it was felt necessary to encourage the use of land resources and new economic activities (Scottish Development Department, 1985).

Whereas the essence of the 1960 policy was maintained, the review allowed for greater flexibility in application of the policy for the varying circumstances that were found across Scotland. It was variations between populations and economic situations, however, which were being acknowledged rather than the range of landscape characters. As a consequence it was accepted that there could be circumstances where new development in the countryside would be allowed.

Government policy on nationally important land use and other planning matters is set out in National Planning Policy Guidelines. Whereas Planning Advice Notes provide advice on good practice and other relevant information for local planning authorities. National Planning Policy Guideline 3 *Land for Housing* (Scottish Office, 1996) confirms that the policy on housing in the countryside, set out in Circular 24/1985: *Development in the Countryside and Green Belts* (Scottish Development Department, 1985), is still appropriate. However, given the need to protect Scotland's rural environment, the Government sought through Planning Advice Note 36: *Siting and Design of New Housing in the Countryside* (Scottish Office, 1991) to encourage a more sympathetic approach to siting and a more widespread adoption of house design which pays greater regard to variations in landscape and building design in Scotland.

10.3 PAN 36: *Siting and Design of New Housing in the Countryside* and its effects

This Planning Advice Note was produced against a background of concern about not only the amount of housing being built in the countryside in the late 1980s, but also about the lack of sensitivity in the siting of new housing, and the introduction of house styles which were more-characteristic of suburban than of rural areas. It explains how planning authorities can enable new housing in the countryside to be developed in harmony with the landscape.

Where a more permissive approach is to be adopted to the location of new housing development, the Planning Advice Note recommends that "the choice of these areas should be carefully researched, with a proper landscape assessment to determine areas which best absorb new housing" (Scottish Office, 1991, p. 3). In preparing new policies for the control of new housing in the countryside, planning authorities are encouraged to take account of

- the scale of new development that would be appropriate in a given area;
- the existing pattern of development and how it fits into the landscape;
- the range of circumstances which exist and the variation in the capacity of different landscapes to absorb development;
- the opportunity that new housing in the countryside can contribute to rather than detract from Scotland's heritage; and
- the nature of local landform and pattern of existing vegetation.

All of these considerations underline the importance of having a greater depth of understanding of landscape character.

Planning Advice Note 36 makes it clear that local plans should interpret, in the light of local circumstances, any presumptions in favour of new housing in the countryside that have been set out in a structure plan. Local plans should also give developers an understanding of the style of development which the planning authority regard as compatible with the character of an area. The Planning Advice Note suggests that a great deal can be learnt by observation of the way in which traditional buildings have been designed and set in the landscape.

The capacity of different types of landscape to absorb development is also discussed in Planning Advice Note 44: *Fitting New Housing Development into the Landscape* (Scottish Office, 1994). The document is complementary to PAN 36 and while the advice is more concerned with sizeable developments on the edge of towns, the recommended approach to site analysis is also relevant for single or small groups of houses in more open countryside.

Research undertaken four years after publication of PAN 36 (Moir *et al.*, 1997) clearly showed that the approach adopted by district planning authorities to housing in the countryside varied throughout Scotland. Constraints on developments in the countryside were being applied throughout much of Lowland Central Scotland, with 'outliers' of control seen elsewhere, for example around Aberdeen. By contrast, in the more remote rural areas including much of Highland Region, rural Grampian, north Angus and the Borders, a more relaxed approach to housing in the countryside appeared to have emerged. This difference might well have been anticipated as it was largely with such remote rural areas in mind that national policy was modified in 1985 to allow a more flexible approach where circumstances merited some relaxation.

The research also indicated that, alongside their policies on the scale and location of housing developments, local planning authorities were using design policies to safeguard the visual amenity of the countryside. Yet the study showed that statutory local plan design policies were brief, generalised or poorly worded. "The local plan appears to be an inadequate mechanism for securing improvements in design standards" (Moir *et al.*, 1997, p. 327). These findings confirmed the Scottish Office advice in PAN 36 that local plans are not the best means of conveying information about siting and design to prospective applicants for planning permission, and it is much better achieved by an illustrated guide. There is a possible danger, however, that design guides might produce a standardised approach to housing. Moir *et al.* (1997) found that relatively few authorities, at that time, had emphasised and subsequently defined what was meant by 'local and regional character'. However, it should be remembered that, in the mid-1990s, there was a limited number of completed landscape assessments available to underpin local plan policy.

Notwithstanding this overall situation, note should be taken of the useful work done by Moray District Council in the early 1990s (Moray District Council, 1993). Recognising that planning applications for housing in the countryside were often controversial, the Council adopted a set of policies and guidelines which were incorporated into the statutory Local Plan. The objective of the policy was to ensure that residential development would be compatible with the preservation of the rural attractiveness of the Plan area. The policy identified three options for prospective developers and set out a series of criteria which must be satisfied if planning approval is to be issued. Interestingly, in the option concerning new build in the open countryside, the policy checklist made it clear that proposals must

- respect existing land form and relate to natural features which provide boundary enclosure and backdrop;
- relate to the traditional pattern of settlement, but avoid a build up of residential development out of character with the area;
- avoid an adverse impact on prime agricultural land, scenery, amenity woodland, trees, natural habitats or sites of scientific importance;
- ensure vehicular access and services are adequate; and
- be designed to reflect site features and relate to architectural styles found in rural Moray.

The policy also recognised five broad landscape types in Moray and gave textural and graphic guidance on the positioning of housing sites in each of these landscapes. Advice is given on both the placement of a house within its site boundaries and its design. The advice is given in graphic representation and is reproduced in Figures 10.1 and 10.2. Emphasis is placed upon the importance of using a rural rather than suburban style of architecture.

The policy and guidelines were reproduced in a separate, supplementary book rather than being produced as part of the local plan. The housing document was reproduced to a very high standard incorporating sketches, plans, diagrams and colour photographs. Publication costs were high and it is now unlikely that the same order of funding will be

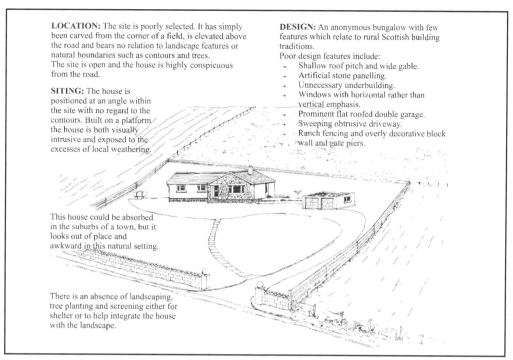

LOCATION: The site is poorly selected. It has simply been carved from the corner of a field, is elevated above the road and bears no relation to landscape features or natural boundaries such as contours and trees. The site is open and the house is highly conspicuous from the road.

SITING: The house is positioned at an angle within the site with no regard to the contours. Built on a platform the house is both visually intrusive and exposed to the excesses of local weathering.

This house could be absorbed in the suburbs of a town, but it looks out of place and awkward in this natural setting.

There is an absence of landscaping, tree planting and screening either for shelter or to help integrate the house with the landscape.

DESIGN: An anonymous bungalow with few features which relate to rural Scottish building traditions.
Poor design features include:
- Shallow roof pitch and wide gable.
- Artificial stone panelling.
- Unnecessary underbuilding.
- Windows with horizontal rather than vertical emphasis.
- Prominent flat roofed double garage.
- Sweeping obtrusive driveway.
- Ranch fencing and overly decorative block wall and gate piers.

Figure 10.1. An example of a planning application that was refused. This siting and design summary was prepared by Moray District Council.

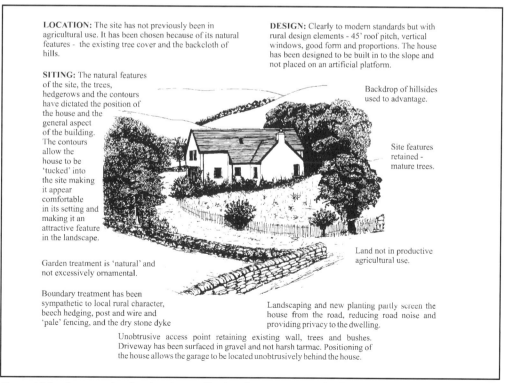

LOCATION: The site has not previously been in agricultural use. It has been chosen because of its natural features - the existing tree cover and the backcloth of hills.

SITING: The natural features of the site, the trees, hedgerows and the contours have dictated the position of the house and the general aspect of the building. The contours allow the house to be 'tucked' into the site making it appear comfortable in its setting and making it an attractive feature in the landscape.

Garden treatment is 'natural' and not excessively ornamental.

Boundary treatment has been sympathetic to local rural character, beech hedging, post and wire and 'pale' fencing, and the dry stone dyke.

DESIGN: Clearly to modern standards but with rural design elements - 45' roof pitch, vertical windows, good form and proportions. The house has been designed to be built in to the slope and not placed on an artificial platform.

Backdrop of hillsides used to advantage.

Site features retained - mature trees.

Land not in productive agricultural use.

Landscaping and new planting partly screen the house from the road, reducing road noise and providing privacy to the dwelling.

Unobtrusive access point retaining existing wall, trees and bushes. Driveway has been surfaced in gravel and not harsh tarmac. Positioning of the house allows the garage to be located unobtrusively behind the house.

Figure 10.2. An example of a planning application that was approved. This siting and design summary was prepared by Moray District Council.

available for a second and updated edition. Nevertheless, Moray Council's Development Control Officers consider the quality of housing proposals in the area has noticeably improved since publication of the guidance. The value of such guidance has been recognised by other planning authorities who have published similar documents (Borders Regional Council, 1993; Perth and Kinross District Council, 1994).

10.4 Application of Landscape Character Assessments by Planning Authorities

Landscape character assessments have recently been completed for the whole of Scotland as a result of the national programme carried out by Scottish Natural Heritage, in partnership with local authorities and other agencies. An examination of published work, together with an unpublished survey of a sample of seven Scottish planning departments, undertaken in mid-1998 indicated that there was a number of strategic or sub-regional planning cases where assessments have played an important role in policy formulation. However, the situation at a more local and detailed site level appeared, particularly in respect of single and small groups of houses in the countryside, to be more mixed.

So far the application of landscape character assessment as part of the process of preparing structure plans appears to be more successful. For example, a detailed assessment of the landscape character of the countryside surrounding St Andrews, including the definition of countryside types, was completed by David Tyldesley & Associates in April 1996 for Fife Regional Council and North East Fife District Council, in association with Scottish Natural Heritage. One of the aims was to the inform preparation of the St Andrews Strategic Study which would identify and assess areas for potential housing development and to provide guidance on how such development might be accommodated. These possible developments concerned the planned future expansion of St Andrews to meet predicted housing land demand rather than isolated housing development in the countryside. Notwithstanding this experience at the strategic level, Fife Council has still to apply landscape character assessments either in local plan policies or to develop more detailed guidelines for development control work in the wider countryside.

Aberdeenshire Council has applied the results of landscape character assessments in both local plan policies and development control decisions. The Consolidated Aberdeenshire Local Plan (Central) contains Policy AC/ENV20 regarding Areas of Regional Landscape Significance and Policy AC/ENV21 concerning Landscape Character. The formulation of both policies draws upon a Countryside Landscape Assessment, carried out by the planning authority, which identified nine character zones throughout the Plan area. These zones are defined on the Main Proposals Map of the Local Plan and a summary of their different characteristics is provided in an appendix. Policy AC/ENV21 states that "there will be a presumption against development being permitted if it could adversely affect the established character of the local landscape in terms of scale, siting, form or design" (Aberdeenshire Council, 1998a, p. 135). The policy lays clear emphasis on the importance of maintaining landscape character. It also states "where it is accepted that a proposal could meet with the established character of the local landscape, the detailed design will only be permitted if it includes associated landscaping appropriate to the surroundings and in scale with the development".

In producing this policy the Council recognised that all of the landscape of Aberdeenshire Central, whether it is 'spectacular' or 'ordinary', is important to someone.

Each of the landscape areas is regarded as important to those who live in them and it is considered important that any new development should enrich the landscape. The purpose of the policy is clearly stated "to ensure that development is responsive to its local landscape setting" (Aberdeenshire Council, 1998a, p. 135). This policy is now being applied in development control work with the full support of the members of the Planning Committee. Refusal of planning consent has been made on the grounds that a "particular development would be conspicuous in the surrounding area and would adversely affect the established character of the local landscape" (Aberdeenshire Council, 1998b, p. 5).

The mid-1998 survey demonstrated that the other six planning authorities had not made as much progress as Aberdeenshire Council; indeed, some had not even begun work to apply landscape character assessments to local plan policy. In part this might be explained by local government reorganisation in April 1996. New unitary planning authorities are likely to have given priority to the review of authority wide strategic policies before moving on to the review of local plans and preparation of detailed, supplementary guidance. However, the survey has shown that some Planning Officers are finding it difficult to apply the broader scale findings of landscape character assessments at the level of detail that is required for local plan work and individual site appraisal. It is also evident that the brief for some landscape assessments has not given sufficient weight to the possibility that they may be used for detailed integration of built development into the countryside. Perhaps there should not be too much surprise at these reactions. The *Landscape Assessment Guidance*, adopted as the standard methodology for landscape character assessment in Scotland by Scottish Natural Heritage, is quite clear that, before any landscape assessment is begun, "the purpose of the exercise should be defined very clearly" (Countryside Commission, 1993, p. 7). It is also quite specific about the importance of deciding, prior to starting the assessment, who it is aimed at, and what level of detail is required. There appears to be a danger that some Planning Officers are trying to develop detailed locational guidelines from assessments which did not include this purpose as a priority but which were intended as an aid to the review of landscape designations or the identification of landscapes sensitive to change.

Difficulty in using landscape character assessments to guide the integration of new housing development into the countryside may also arise from the way in which the Countryside Commission's guidelines have been applied. The guidelines are "concerned primarily with broad, area-wide landscape assessments, rather than with detailed, site-specific assessments" (Countryside Commission, 1993, p. 3). Neither does the section explaining the techniques of assessment give as much emphasis to assessment of settlement pattern as it does to landform and land cover analysis. It is evident, therefore, that the value of landscape character assessment for the integration of new housing in the countryside is very much greater when the assessment is specifically prepared with that purpose clearly in mind.

The application of landscape character assessment in development control work must take account of the procedural context in which planning decisions are taken. Planning authorities are under increasing pressure to meet targets set by the Scottish Office for the percentage of planning applications determined within eight weeks of submission. To meet these targets, Development Control Officers have little time to negotiate with developers any revisions to the siting, design and landscaping aspects of proposals. As a consequence, guidelines concerning landscape character as an influence upon the formulation of acceptable planning applications must be presented in clear and precise language so that applicants can understand what is expected by the planning authority. One estimate from one rural planning authority indicates that the clear majority of planning applications for new houses in the countryside have been prepared by draughtsmen with no design qualification. Consequently raising awareness of landscape character and the importance of securing development that will 'fit' into the landscape is likely to be an important, yet lengthy, process.

The importance of securing this 'fit' is shown by means of two photographs. Plate 17 shows an example of a poor site selection, the house being positioned on a low skyline. The house is visually prominent and is out of character with the surrounding open scrubland. In contrast, Plate 18 shows a much more satisfactory house site. The development is visually integrated within the mixed, enclosed landscape and is backed by rising ground and established trees.

10.5 Conclusions

Recent government publications recognise that many of the rural areas of Scotland most in need of economic and social development are places where the countryside is particularly valuable and sensitive (Scottish Office, 1998 a & b), as evidenced by the quotation "Scotland's natural heritage is one of its most important assets" (Scottish Office, 1998a, p. 3). There are landscapes of international renown which attract visitors from around the world but which are also important to local communities for economic and social benefits. It is important, therefore, that new development is located in sympathy with the landscape and is designed and constructed in a manner that respects local landscape character and building styles. New housing is the major form of development in rural Scotland and it is often visually prominent (Scottish Office, 1998a). In order to maintain the visual character of the countryside, increasing care is required in the siting and design of new houses.

In England there is a government requirement, set out in Planning Policy Guidance 7 (Department of the Environment, 1997), to replace existing local landscape designations in development plans with a landscape character-based approach. The objective is to sustain the distinctiveness, the key features and the characteristics of all landscapes irrespective of judgements of scenic beauty. This approach has been reinforced by the introduction of *Village Design* and *Countryside Design Summaries* (Countryside Commission, 1996 a,b). These methods of working have been advocated by the Countryside Commission and the former particularly involves the local community in the formulation of statements about new housing development. The Scottish Office (1998a) consultation asked whether this experience had anything to offer in Scotland.

As a result of the national programme of Landscape Character Assessment promoted by Scottish Natural Heritage (see Hughes and Buchan, this volume), there is now available a wealth of information about the range and variety of Scottish landscapes. The question arises, therefore, as to how this information might be used by town and country planners, particularly in respect of new housing in the countryside. Experience reviewed in this paper indicates that there is an established positive role for landscape character assessment at the strategic planning level. However, experience at local plan level is mixed and generally less certain. The Consolidated Aberdeenshire Local Plans 1998 and the Aberdeenshire

Council's consequential development control work both show what is possible. On the other hand, some planning authorities have still to apply landscape character assessment to local plans and development guidelines. Accordingly, it is suggested that Scottish Natural Heritage might undertake the following four actions.

- In partnership with planning authorities, identify areas where there is particular pressure for new housing in the countryside and ensure that landscape character assessments are sufficiently detailed and specific to inform the formulation of local plan policy and detailed supplementary planning guidance.
- Issue guidance on the integrated use of landscape character assessments and sub-regional studies of building design and construction so as to achieve appropriate settlement patterns. The availability of such guidance could save considerable planning authority staff time that is currently spent explaining to applicants the landscape integration of new development.
- Put in place a series of training events for local authority planners regarding the application of landscape character assessments in the whole range of statutory planning work.
- Consider mounting a pilot project, in partnership with the planning authority, landowners and local community, to identify an appropriate scale and pattern of new housing that is required to meet the needs of local people, yet still respects, the landscape resource.

The availability of landscape character assessments represents a potentially important new tool for planning and controlling new housing in the countryside. Government supports an increasingly flexible approach on this subject in development plans and development control, particularly in remote rural areas where new housing development is seen to be an important means of sustaining the local economy. There is evidence that elected members serving on planning committees want to conserve Scotland's landscape resource, but they also wish to allow appropriate development. Landscape character assessment should be seen as a means of meeting both objectives.

References

Aberdeenshire Council (1998a). *Consolidated Aberdeenshire Local Plans 1998*. Aberdeenshire Council, Aberdeen.

Aberdeenshire Council (1998b). *Garioch Area Committee Report - 28 July 1998*. Aberdeenshire Council, Aberdeen.

Angus District Council (1980). *Advice Note 5: Houses in the Open Countryside*. Angus District Council, Forfar.

Borders Regional Council (1993). *New Housing in the Borders Countryside*. Borders Regional Council, Newtown St Boswells.

Cherry, G. (1975). *National Parks and Recreation in the Countryside. Environmental Planning 1939-1969, 2*. HMSO, London.

Countryside Commission (1993). *Landscape Assessment Guidance, CCP 423*. Countryside Commission, Cheltenham.

Countryside Commission (1996a). *Village Design: Making Local Character Count in New Development, CCP501*. Countryside Commission, Cheltenham.

Countryside Commission (1996b). *Countryside Design Summaries - Achieving Quality in Countryside Design, CCP502.* Countryside Commission, Cheltenham.

Department of the Environment (1997). *Planning Policy Guidance 7: The Countryside - environmental quality and economic and social development.* Department of the Environment, London.

Department of Health for Scotland (1960). *Circular 40/1960: New Houses in the Country.* Department of Health for Scotland, Edinburgh.

Highland Regional Council (1977). *Development Control Policies: Housing.* Highland Regional Council, Inverness.

Moir, J., Rice, D. and Watt, A. (1997). Visual amenity and housing in the countryside - Scottish local planning authority approaches. *Land Use Policy,* **14**, 325-330.

Moray District Council (1993). *Moray District Local Plan 1993-98: Housing in the Countryside.* Moray District Council, Elgin.

Perth and Kinross District Council (1994). *Guidance on the Siting and Design of Houses in Rural Areas.* Perth and Kinross District Council, Perth.

Scottish Development Department (1985). *Circular No 24/1985: Development in the Countryside and Green Belts.* Scottish Development Department, Edinburgh.

Scottish Development Department (1987). *National Planning Guidelines: Agricultural Land.* Scottish Development Department, Edinburgh.

Scottish Office (1991). *Planning Advice Note 36: Siting and Design of New Housing in the Countryside.* Scottish Office and HMSO, Edinburgh.

Scottish Office (1994). *Planning Advice Note 44: Fitting New Housing Development into the Landscape.* Scottish Office Environment Department, Edinburgh.

Scottish Office (1996). *NPPG3 (Revised) Land for Housing.* Scottish Office Development Department, Edinburgh.

Scottish Office (1998a). *Investing in Quality: Improving the Design of New Housing in the Scottish Countryside - A Consultation Paper.* Stationery Office, Edinburgh.

Scottish Office (1998b). *Draft NPPG Rural Development.* Scottish Office Development Department, Edinburgh.

PART THREE

Techniques and Models

11 USING AERIAL PHOTOGRAPHY IN STATIC AND DYNAMIC LANDSCAPE VISUALIZATION

D.R. Miller

Summary

1.   Aerial photography can provide a core input to enable the visualization of landscape.
2.   Digital elevation models of a high accuracy and horizontal resolution can provide visualizations of robust quality.
3.   Spatially referenced video fly-throughs can be used to present changes in landscape.
4.   To be useful for interpreting detailed changes in land cover features the relative horizontal accuracy of aerial photography for multiple dates should be equivalent to the pixel size of the dataset with the lowest horizontal resolution.

11.1 Introduction

Aerial photography provides an important source of data for the interpretation of landscape features (Howard, 1970). Extensive coverage exists for almost all of the United Kingdom since 1946, with the following three national surveys

- Royal Air Force reconnaissance photography between 1946 and 1957;
- Ordnance Survey photography for use in the national large scale mapping programme; and
- the national land cover survey undertaken by the Scottish Office in the late 1980s.

There are also numerous targeted flight sorties undertaken for both government agencies and commercial companies. Therefore, aerial photography provides a source of temporal data for most areas of the United Kingdom and this chapter presents an approach to identifying and visualizing changes in the landscape through time. The objectives are

- to describe the data requirements for identifying changes in landscape features;
- to demonstrate a method for processing and presenting landscapes at different periods in time, including the use of digital data derived from historical aerial photography; and
- to construct video fly-throughs of the landscape to show the changes in landscape.

11.2 Background

Appraisals of the type and extent of the natural resources of an area are required for effective decision making in environmental management from local to national levels (Aspinall *et al.,*

1992; Orland, 1992; Zewe and Koglin, 1995). Appraisal for a single date may provide valuable information for some applications but only with information for additional years can assessments be made of natural or human-induced change in the landscape (Hester *et al.*, 1996). Typical uses of information on changes in landscape include the monitoring or surveillance of landscape to assess the effectiveness of policy decisions, such as changes within areas classed as priority natural habitats in the EC Habitats Directive (Anon., 1992) and the assessment of the impacts of change, such as that of afforestation (Anon., 1988).

The identification, measurement and communication of changes in landscape fall into five broad categories: textual description, statistical summaries, maps of change, photographs (terrestrial and aerial) and three-dimensional (3-D) computer models. Often it is a combination of different approaches which provides the most comprehensive impression of change. The four principal issues, common to all approaches, are: data capture (including census and sampling techniques); data processing for ensuring consistency in classification, geometry, reference system and level of detail; interpretation and measurement of features or classes; and presentation of the results (Faust, 1995; P. Berger, pers. comm.).

This chapter describes a method for processing historical and contemporary aerial photography as basic data for the measurement and visualization of changes in the landscape using 3-D modelling. It also discusses the issues associated with the preparation of the aerial photography and the requirements for digital elevation models. Finally, the chapter describes and discusses how video compilations of landscape views can be used to present landscape change. The examples used are from two areas that have undergone contrasting land cover changes over the past three or four decades and for which aerial photography was available. The first is Glenfeshie in the western Cairngorm mountains of Scotland; the second is the Cwm Berwyn forest in mid-Wales (Figure 11.1).

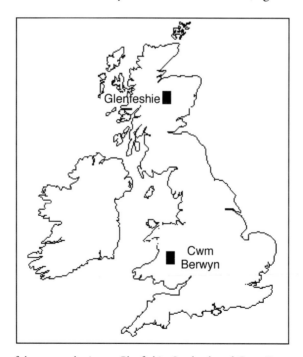

Figure 11.1. Location of the two study sites at Glenfeshie, Scotland, and Cwm Berwyn, Wales.

11.3 Methods and Data Processing

The core datasets used as inputs for the interpretation and presentation of landscape and landscape changes are

- digital elevation models (DEMs), which are digital representations of the elevation of locations on the earth's surface (McDonnell and Kemp, 1995); and
- orthophotographs, which are rectified aerial photographs that show objects in their true planimetric positions (Baltsavias, 1996). They are geometrically equivalent to conventional line maps and thus can be used to scale off distances, angles, positions and areas.

The significance of these with respect to the two study areas is that they each contain steeply sloping land which would make direct comparisons of small landscape features, between the dates of photography, unfeasible.

The aerial photographs for each area were scanned at a resolution of 400 dots per inch and then processed using the ERDAS OrthoMax photogrammetric package (Anon., 1997). Creation of the orthophotograph involves five steps, namely

- setting the camera calibration characteristics and measuring the locations of the fiducial marks on the photographs;
- measuring the locations of tie points and ground control points for all of the stereo models that cover the area;
- derivation of the DEM using the stereo aerial photographs and a numerical model formed from the camera characteristics, tie and ground control points;
- ortho-correcting one of the overlapping aerial photographs with respect to the DEM using the relevant stereo-model; and
- mosaicing of adjacent orthophotographs to provide cover for the entire area of interest (Miller *et al.*, 1996).

Creation of the datasets from the 1946 (Glenfeshie) and 1957 (Cwm Berwyn) photography required camera calibration information to be obtained from the Joint Air Reconnaissance Information Centre, but no details were available to identify which camera was used in the photographic sorties. However, low oblique photographic data cover a large proportion of the United Kingdom and provide the most comprehensive and valuable source of historical data for interpreting land cover and landscape change during the period since 1945. Therefore, the quality of orthophotographs derived from such photography was tested in the two study sites so as to assess their suitability for use with more recent aerial photographs.

Ground control data were obtained from a combination of differential Global Positioning System data and Ordnance Survey Land Line (1:2,500 and 1:10,000) digital maps. These sources provided horizontal and vertical ground control data to accuracies of 1.3 m and less than 1 m (Anon., 1996) respectively. Controls of the most recent photography (Glenfeshie, 1988; Cwm Berwyn, 1995) were based upon the selection of field and map features for reference and check points. There was a limited number of recognisable map features in the steeper and higher altitude land which characterizes the areas. Therefore, the advantage of processing the most recent photography first was that features on the resultant orthophotograph could be used as control for the processing of the older dates of photography. This is because the rock and vegetation features, interpretable from the photography, are not mapped by the Ordnance Survey but can be used as control points.

The digital photogrammetric software was used to derive high resolution DEMs (1 × 1 m or 2 × 2 m) from each set of stereo pairs of aerial photographs. Adjacent DEMs were mosaiced together to provide complete coverage for each study area. However, one significant limitation to the effective use of the photographs, particularly those which deteriorated over time (such as the 1946 and 1957 imagery listed in Table 11.1), is the difficulty of joining a series of separate orthophotographs in order to create a larger area of orthophotographic coverage (i.e. an orthophotographic mosaic). With the older, historical, aerial photography the considerable differences in photographic exposure between flight lines can limit the successful interpretation of that imagery. Where possible, therefore, the visual quality of the orthophotographic mosaics was maintained by minimizing the possible disjunction in appearance at their edges either by using the same photograph for two adjacent photogrammetric models, thus ensuring that the common boundary of the models was 'seamless', or, where a choice exists, selecting those aerial photographs which had the most consistent tone, colour and reflective appearance.

A distinction can be made between the value of pictorial quality in its own right and the quality of the photographic image required for the interpretation of change. The creation of the mosaics can introduce 'a false boundary' by the choice of the 'cut' along which to trim the photographs; that is, at the join of two edges the different tones, textures or colours may be credible in a landscape view but misleading in the interpretation of vegetation type.

Table 11.1. Details of the photographs used in each study site and the orthophotographs produced. Absolute accuracy is the root mean square (RMS) of the co-ordinates of check points identified on the aerial photography against ground control points; relative accuracy is the RMS of the co-ordinates of check points identified on one set of orthophotography against those of another set of orthophotography available for the same site.

Area	Date	Scale	Number of photographs	Horizontal resolution (m)	Absolute accuracy (m)	Relative accuracy (m)
Glenfeshie	1946	1:20,000	6	2	2.9	1.7
	1988	1:24,000	9	1	2.1	
Cwm Berwyn	1957	1:20,000	2	2	2.3	1.4
	1975	1:26,000	4	2	2.2	
	1992	1:10,000	5	1	1.1	
	1995	1:10,000	5	1	0.9	

Table 11.2. Summary of resolution and accuracy of the digital elevation models (DEMs) derived for each study area. In most cases the resolution and accuracy of the DEM is the same as those of the orthophotographs.

Area	Date	Horizontal resolution (m)	Absolute accuracy (m)
Glenfeshie	1988	2	2.1
Cwm Berwyn	1975	2	1.5
	1992	1	1.1
	1995	1	0.9

11.4 Dataset Derivation

The data in Tables 11.1 and 11.2 summarize the horizontal resolution (i.e. the level of detail of the spatial data) and the accuracy of the orthophotographs and DEMs respectively for the more recent photography of Glenfeshie in 1988 and Cwm Berwyn in 1995. The accuracy of the output imagery has been reported in terms of two estimates of geometric accuracy. These are 'absolute accuracy', which is the root mean square (RMS) of the co-ordinates of check points identified on the aerial photography against ground control points; and 'relative accuracy', which is the RMS of the co-ordinates of checkpoints identified on one set of aerial photography against those on another set of photography. Checks on the accuracy of the DEMs are summarised in Table 11.2. These results show an absolute accuracy of between ±2.1 m and ±0.9 m in elevation for the DEMs used. The absolute error in the DEM data results from inaccuracies in the ground control co-ordinates and the photogrammetric model. Mis-registration between adjacent DEMs may be due to different representations of the terrain and the surface features on each DEM and errors in the joining of the DEMs, leading to the introduction of discontinuities in the data over which the orthophotographs will be draped.

The results obtained for the orthophotograph production show a maximum absolute error of ±2.9 m for the Glenfeshie data for 1946 and the lowest absolute error of ±0.9 m for the Cwm Berwyn data of 1995. These results reflect the scale and quality of the photography at the time and the availability of ground control data. However, for identifying changes in land cover features between different dates of photography, the relative accuracy is more important and this has been calculated as being between ±1.7 m and ±1.4 m for these two study sites. Therefore when overlaying multiple dates of orthophotography the error in co-registration is less than the resolution of the coarsest dataset.

11.5 Presentation and Identification of Landscape Changes

11.5.1 Presentation of Landscape Change

To aid in the presentation and communication of the changes in landscape over the 40 years, fly-throughs and video sequences were created for replay on a PC, using QuickTime format and an associated player. This format and software enables the use of the video imagery on commonly available computing platforms that are easily transported or the use of CD-ROMs for wider dissemination. The preparation of such an output is in three stages, namely

- the creation of a flight path, chosen to highlight those areas or features that one wishes to present to the viewer;
- the recording of the individual perspective views into a video sequence; and
- the editing of video sequences to create a single video file from which different dates of imagery can be presented.

11.5.1.1 Flight Path Creation

The software used in this study allows the digitizing of a flight path using the orthophotograph as a reference image. The flight paths require some editing to produce a 'smooth' movement across the landscape.

To select a path which highlighted certain locations and topographic features it was also found necessary to create sections of flight path with different parameter settings, such as the observer's angle of view (horizontal and vertical), the flying height above the ground and the speed of travel during a flight. This was particularly the case in the Glenfeshie study area, for which a movie sequence of a flight across the landscape in 1946 and in 1988 was created. Certain locations were visited where either change was dramatic over the time period or key landscape features were to be highlighted. Figure 11.2 illustrates a sequence of views taken from the fly-through of Glenfeshie for 1988.

For Cwm Berwyn, flight paths were created to enable each part of the 1995 forest to be viewed at each of the four dates of available photography. In this case, the routing of the flight path was based upon the 1995 photography to ensure that there was comprehensive coverage of the forest. The video sequences were then compiled using the same flight path but different orthophotographs and DEMs. The success of the flights were assessed in terms of

- the successful viewing of all the points intended;
- producing an image size that was sufficient to provide either context or detail to the perspective view;
- achieving a movement across the landscape which was not too rapid to identify features; and
- avoiding abrupt changes between scenes.

When testing the flights, the data were sub-sampled by factors of 10 or 15 for both the Glenfeshie and Cwm Berwyn sites. This was because the size of these datasets, when displayed at full resolution, became impractical to traverse sufficiently rapidly.

11.5.1.2 Sequence Editing

The final outputs for the presentation of changes in the landscape are sequences of fly-throughs across different dates of imagery. Videos comprising sequences derived for different years have the potential to convey the impression of where and to what extent there have been changes in land cover and use. Selection of suitable imagery and interlacing of

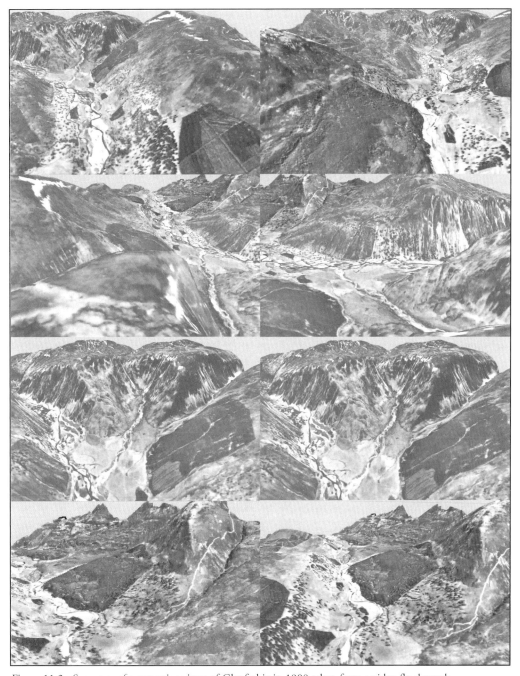

Figure 11.2. Sequence of perspective views of Glenfeshie in 1988 taken from a video fly-through.

sequences can enable changes over different periods of time to be emphasised. However, to produce effective videos, there has to be coherence between different sequences so that the viewer is persuaded that he is seeing the same scene at different dates and is not distracted by discontinuities and differences in the resolution, scale, image size or colour of each sequence. This necessitates the spatial accuracy obtained in the earlier processing of the datasets.

The editing of multiple sequences of videos requires some consideration of how the output will appear at the merging of those sequences. If the input sequences are sufficiently well matched in their characteristics, an abrupt switch between sequences of different dates may be effective. For example, where the change in land cover is dramatic, such as the felling of a forest stand, this approach will enhance the impact of the change.

The sets of videos can then be edited to compile a video-montage from data of more than one year. For example, the data have been edited such that the viewer sees the landscape in 1946 when flying in one, approximately, circular flight path and then tours the same flight path in 1988. However, if the characteristics of the data are so different that the viewer would find them distracting, other approaches require to be considered. One option adopted has been to make the end point of the flight across one date of imagery the same as the start point of a flight across that of another date. This provides video sequences of the same view at different points in time. The transition between imagery has then been hidden by the progressive 'fogging' of the end of the first sequence and the reverse for the start of the second sequence, which has the effect of obscuring any discontinuities in the montaged sequence.

11.5.2 Identification of Landscape Change

The significance of this combination of levels of detail and accuracy is that land cover features can be identified and measured, for example, to the level of the crowns of individual

(a)   (b)

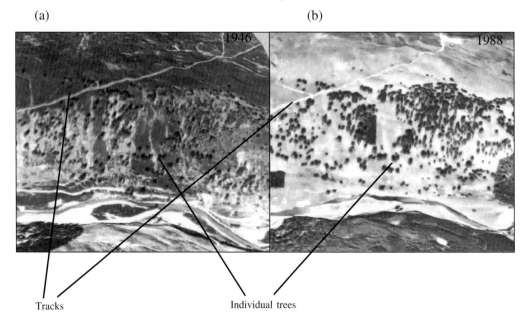

Figure 11.3. Perspective views of a part of Glenfeshie using aerial photography for (a) 1946 and (b) 1988.

trees, as illustrated in Figure 11.3 for Glenfeshie. The Glenfeshie area exhibited several gross, human-induced, changes in land cover such as afforestation (plantations) and deforestation (native woodland) and the introduction of new hill tracks. Other changes in the semi-natural environment, such as transitions within the heather moorland, or between it and the grassland communities, have been studied using the same sets of aerial photographs and reported by Hester *et al.* (1996).

The tracks created for the extraction of the timber are visible. The presence of many individual trees which were not felled is highlighted by the effect of their shadows being 'stretched' by the draping of the imagery across the sloping hillside as represented in the DEM. The woodland has been replanted around the existing pines, the crowns of which are still visible on the photographs.

In the Cwm Berwyn example, the degree of afforestation was more dramatic than for Glenfeshie and the stage in the growth of the forest was such that in 1992 the canopy was closed and in 1995, several areas had been felled. Figure 11.4 shows the landscape perspective views across the forest for 1957, 1975, 1992 and 1995.

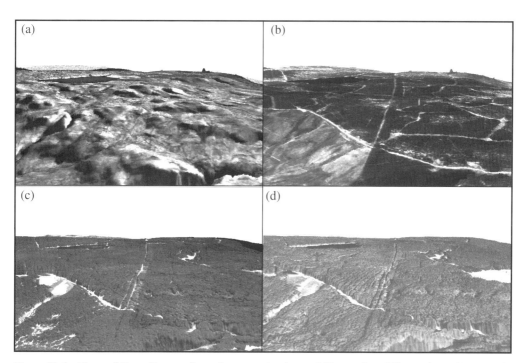

Figure 11.4. Views of Cwm Berwyn forest from pre-planting to felling at four different dates. (a) 1957; (b) 1975; (c) 1992; and (d) 1995.

In the pre-planting period in 1957 the land use was predominantly one of agriculture. The land around the lake and on the hilltops was used for sheep grazing and there were no trees in the area. On the lower slopes, the land was enclosed and used for improved pasture and forage crops. By 1975, most of the area had been planted, with concentrations on the hilltops and the land around the lake. In 1992, the trees were approximately 32 years old and the management of some areas had included line thinnings and the felling of a few stands, which are visible in Figure 11.4c. By then, the canopy had closed across much of the forest. However, by 1995, several large areas had been felled and the number and extent of the management units, which had been thinned, had increased.

These changes in the forest were very evident from viewing the video 'tours' of the area because video editing was used to compile a sequence which accentuated the visual effects of the felling. This sequence used an extract of the Cwm Berwyn flight line in which the viewer was taken across a valley towards the forest, as seen in 1992. A second sequence was created from the same extract of the flight but played in reverse and using the data for 1995 in which a forest stand had been felled. The two sequences were then edited together.

Copies of the video fly-throughs mentioned in this paper are available at the world wide web address http://www.mluri.sari.ac.uk/landscape/.

11.6 Conclusions

The principal data input described in this chapter is aerial photography. It has been used to derive digital elevation models and orthophotographs for different years. The results show an accuracy that is similar to, or better than, that reported by the Ordnance Survey for their height datasets (Anon., 1996), and similar to the 1:10,000 scale of map data.

The value of the historical aerial photographs in visualizing landscapes as they may have appeared in the past lies, in part, in the flexibility they provide to observe the terrain where no terrestrial photographs have been taken. Maintaining sequences of aerial photographs for areas of particular interest allows comparisons to be made between landscape views, taken at different times, so that information interpreted from one year's photography may be compared to that of another year. Areas of planting and clear-fell, for example, can be viewed with respect to the underlying topography and, with interpretations of vegetation types, from a historic perspective. Naturally, such views may be derived for locations and in directions selected by the user; these may include points that are currently inaccessible on the ground but which are of potential future importance. If the focus of interest changes, the landscape study is not restricted as it would be if selected historic, terrestrial photographs were used.

The presentation of geographic data using movie recordings also confers several advantages that may become increasingly important in the future. These advantages include simplicity in the guided presentation of landscape scenes when derived from three-dimensional models. However, they also provide an example of one means of communicating the visual impacts of change to different audiences for purposes of advocacy, or in describing and reporting upon issues of landscape change where the paper medium might be inadequate.

With the near-universal availability and use of GIS packages within the relevant authorities, there is the potential to develop the relationship between the planning instruments concerned with proposals for future changes in the landscape and the evaluation of similar proposals on the historical changes in landscape. Work in related areas will enable closer links between the visualization and presentation of changes in landscape together with experience of, and preference for, landscapes. The coupling of all of these tools for landscape analysis, and the possibility of using high-resolution satellite imagery, provides an aid for the assessment of landscape which could enhance the quality of decision-making with regard to the potential impacts of change.

Acknowledgements

I wish to acknowledge the Scottish Office Agriculture, Environment and Fisheries Department for the funding of this project, and ERDAS (UK) Ltd and Silicon Graphics Ltd for the use of the software and hardware. Thanks are also due to Chris Quine at the Forestry Commission for access to the aerial photography for Wales; to the Royal Commission for the Ancient and Historic Monuments of Scotland for some of the aerial photography for Scotland; and to Jeff Maxwell, Dick Birnie and two anonymous referees for their contributions to the text.

References

Anonymous 1988. *Environmental Assessment (Afforestation) Regulations.* Edinburgh, Forestry Commission.

Anonymous 1992. *Conservation of Natural Habitats and of Wild Flora and Fauna. Council Directive 92/47/EEC.* Brussels, European Commission.

Anonymous 1996. *Profile DEM User's Manual.* Southampton, Ordnance Survey.

Anonymous 1997. *OrthoMax User's Manual.* Atlanta, ERDAS Inc.

Aspinall, R.J., Miller, D.R. and Birnie, R.V. 1992. GIS for rural land use planning. *Applied Geography,* **13**, 54-66.

Baltsavias, E.P. 1996. Digital ortho-images: a powerful tool for the extraction of spatial and geo-information. *ISPRS Journal of Photogrammetry and Remote Sensing,* **51**, 63-77.

Faust, N.L. 1995. The virtual reality of GIS. *Environment and Planning B: Planning and Design,* **22**, 257-268.

Hester, A.J., Miller, D.R. and Towers, W. 1996. Landscape-scale vegetation change in the Cairngorms, Scotland, 1946-1988: implications for land management. *Biological Conservation,* **77**, 41-51.

Howard, J.A. 1970. *Aerial-photo Ecology.* London, Faber.

McDonnell, R.A. and Kemp, K.K. 1995. *International GIS Dictionary.* Cambridge, Geoinformation International.

Miller, D.R., Quine, C.P. and Broadgate, M.B. 1996. The application of digital photogrammetry for monitoring forest stands. In Kennedy, P. and Folving, S. (eds) *Application of Remote Sensing in European Forest Monitoring.* Luxembourg, European Commission, pp. 57-68.

Orland, B. 1992. Data visualization techniques in environmental management. *Landscape and Urban Planning,* **21**, 237-244.

Zewe, R. and Koglin, H.-J. 1995. A method for the visual assessment of overhead lines. *Computers and Graphics,* **19**, 97-1.

12 MAPPING REMOTE AREAS USING GIS

Steve Carver and Steffen Fritz

Summary

1.   Remoteness in the context of landscape character is more than just a simple measure of linear distance to the nearest road.
2.   In Scotland, in particular, remoteness is closely linked to the idea of 'the long walk in' and is strongly influenced by indices of landscape character such as terrain, water features, intervisibility, cultural history, land cover and land management.
3.   While GIS is particularly good at generating buffer zones and distance surfaces, careful thought needs to be given to how these capabilities are brought to bear in defining remote areas. This chapter describes an approach to remoteness mapping that is based on incorporating GIS-based accessibility models and landscape data to derive remoteness surfaces that take different definitions of 'remote' into account. The resulting surfaces are discussed in the context of other indices of landscape character.
4.   Possible uses of remoteness surfaces are demonstrated. These include wild land mapping and zoning of protected areas. Examples from the Cairngorm area are given.

12.1 Introduction

One of the key aspects of landscape character attributable to many parts of the Scottish Highlands is their remoteness. Remoteness is not, however, something that is easy to measure, nor indeed describe, in generic terms. Yet it is an essential element of some of the wildest parts of the Highlands. The physical nature and geographical location of these landscapes often instils a feeling of remoteness.

Despite changes in land ownership and management, and the effect they have had on the contemporary Highland landscape, it is the physical influences of landscape character on remoteness that are the focus of this chapter. In particular, the chapter draws attention to how Geographical Information Systems (GIS) and spatial models may be used together to map remoteness as a function of the time taken to walk into roadless areas.

With the significant rise in outdoor recreation seen during the last 40 years, and substantial increases predicted over the next few decades (Ewert and Hollenhorst, 1997), new techniques are required to help manage the landscape resource and in particular those areas beyond mechanised access. Remoteness maps are seen as an essential element in the evaluation of wild areas and in the development of policy aimed at their protection.

12.2 Remoteness as an index of landscape character

12.2.1 Definitions of remote

Remoteness may be defined geographically as the linear distance from one point in space to another point in space. This model of remoteness makes the assumption of equal speed and cost of travel in all directions. This does not hold true in the real world as differences in transport networks, terrain and other variables make such a model too simplistic. Another, more useful, definition of remoteness is the time taken to travel between the origin and destination. Using time of travel as a measure of remoteness requires that a number of geographical factors other than simple linear distance are considered. These include the various cost or push (benefit) factors that not only influence the ease of travel in a particular direction, but also the route chosen. 'Cost' factors have a negative effect on calculated travel times in that some cost beyond the effects of linear distance is applied in travelling between two points on a terrain surface. Typical cost factors are slope (when travelling uphill), headwinds, thick vegetation and difficult ground conditions. 'Push' factors have a positive effect on the speed of travel in that some additional force is applied in the direction of travel. Typical examples are slope (when travelling downhill) and tailwinds. It is noted here that certain cost/push factors, especially slope and wind direction, are strongly dependent on direction of travel.

There are several geographical factors that are likely to influence pedestrian off-road access. These include

- terrain variables such as slope and altitude;
- ground cover variables such as height and thickness of vegetation;
- presence of tracks and paths as these often constitute the route of least resistance;
- prevailing weather conditions, including wind speed, direction, visibility and temperature;
- barrier features such as lochs, crags, unfordable rivers and private land; and
- hydrological conditions, including ground wetness, stream levels and snow cover.

All of these variables may be included in a spatial model to estimate remoteness, given appropriate datasets to work with. It is necessary to distinguish between definitions of remoteness for both mechanised and non-mechanised modes of transport, since different cost/push factors assume very different levels of significance within each category. While much work has been carried out on modelling accessibility along existing road/rail systems using network analysis (Sedgewick, 1984; Douglas, 1993), comparatively little work has been done on off-road access. Although some research has been carried out on all-terrain vehicle access for military applications (Cuddy, *et al.*, 1996), the work focusing on non-mechanised (pedestrian) off-road access remains sparse.

A third potential definition of remoteness is that of perceived remoteness. The individual fitness level of the walker in question is a non-geographical variable that can influence perceived remoteness. The fitter the individual the greater the distances that can be covered in the same time. The physical effects of some of the above variables such as slope, ground cover and weather will have a lesser effect in fit individuals. Nonetheless, geography remains an important factor governing tiredness, as a walker's speed declines with time and distance covered.

Other geographical variables that might have a significant role to play in measuring perceived remoteness are the presence or absence of obvious human features in the landscape, such as old crofts or sheilings, dams, pylons, deer fences and plantation forests. These artificial features can significantly detract from the feeling of remoteness, especially when located deep within roadless areas. One variable that may have a significant effect in influencing the perception of remoteness is intervisibility. If an individual cannot see a nearby settlement or structure as a result of topographic shielding or prevailing weather conditions, then the feeling of remoteness may be enhanced.

While perceived remoteness is important in studies of landscape character, this chapter focuses the use of GIS-based models to map physical remoteness for roadless areas defined as the time taken to walk from the nearest road, taking the physical characteristics of the landscape into account. However, the role of remoteness, both physical and perceived, is recognised as an important aspect of landscape character.

12.2.2 *The role of remoteness in landscape character*

Remoteness plays a significant role in determining people's perceptions of landscape character. In Scotland, in particular, remoteness is closely linked to the idea of 'the long walk in'. In this context, remoteness is defined by the least number of hours it takes to reach a particular destination from any origin (usually a road or car park).

Many people also associate remoteness with wilderness, sometimes with scant regard for land use and landscape histories, substituting inaccessibility for natural ecosystems in their eagerness for a 'wilderness' experience. Remoteness is, from this perspective, best referred to as a *perceived* variable in that while it may have little or no biophysical effect beyond its geographical influence on the history of human use of the landscape, it does have a marked effect on how people feel about the landscape setting. Experiential values play an important role here, in that people who have visited truly wild and remote locations, in places such as Antarctica or the Amazon rainforest, will clearly hold different views about the remoteness of Highland landscapes compared to those who have never set foot outside of a city. Nash (1982, p.1) succinctly illustrated this dilemma with his statement *"One man's wilderness is another's road-side picnic ground"*.

Whereas there is no true wilderness left in Scotland today, large areas of the Highlands engender a feeling of remoteness that many people associate with wilderness. Wilderness as an entity is notoriously difficult to define. Leopold (1921), Nash (1982), Hendee *et al.* (1990) and Oelschlaeger (1991) have all attempted academic definitions, while formal definitions such as that given by the US Wilderness Act (1964) have been written for legislative purposes and are in active use. Most of these definitions stress the natural state of the environment, the absence of human habitation and the lack of other human related influences and impacts. The Outdoor Recreation Resources Review Committee (1962, p.34) defined wilderness as areas over 100,000 acres *"containing no roads usable by the public"* and showing *"no significant ecological disturbance from on-site human activity"*. Applying this definition of wilderness to Scotland would merely result in a blank map. True, pristine wilderness does not exist any more in this part of the world.

The landscape of the Scottish Highlands does, however, engender a feeling of 'wildness' as distinct from 'wilderness' in the strictest sense. This is attributable to the rugged nature of the physical landscape, the predominance of semi-natural vegetation, the lack of people and human artefacts, the presence of wildlife and geographical remoteness. Such parts of Scotland's landscape are referred to here and elsewhere as 'wild land' (Fenton, 1996). Being able to map remoteness accurately by taking the associated geographical variables into account could prove to be of great value to studies of wild land and landscape character. It is possible to estimate for any location within a roadless area the time it would take to walk to that location from a road or car park, i.e. "how long is the long walk in?". Calculating the time taken to walk to all points in a roadless area from the nearest access point, it is possible to generate a remoteness surface for the whole area of interest. This can be used to help define policy boundaries, zone recreation areas, map landscape character and even wilderness quality indices (Fritz and Carver, 2000).

12.3 Mapping accessibility and remoteness using GIS

12.3.1 GIS-based accessibility models

Simple GIS-based access models often start by assuming equal ease of travel in all directions. When considering accessibility from point, line or area origin features, the simplest solution is to draw buffer zones of a set width around these to define zones of equal accessibility. An enhancement to the buffering approach is to dispense with the set zone widths and calculate the absolute distance as an isotropic surface. This surface shows the linear distance (proximity) for each point on the surface (destination) to the nearest access feature (origin). These simple buffering and proximity approaches are shown in Figure 12.1 for the Cairngorm area, where the roads surrounding the plateau are used as linear access features.

More complex approaches employ network analysis techniques that take into account the effect of road and other transport networks on accessibility calculations. These models calculate travel times through the network according to impedance values that have been set for each link (length of road) based on factors such as gradient, speed limit and number of carriage ways. Once the point on the network nearest to the destination has been reached, the assumption of equal ease of travel in all directions then commonly applies as the individual alights from their vehicle and proceeds on foot to their destination.

Some GIS models employ cost/push factors in calculating accessibility surfaces not constrained to transport networks (e.g. Dana Tomlin's MAP software, see Burrough, 1983). In the context of off-road pedestrian travel, terrain variables such as gradient may be applied to give better estimates of access times. By incorporating relevant cost/push factors into pedestrian specific access models within a GIS, better estimates of remoteness for a given roadless area may be obtained.

12.3.2 Incorporating additional factors influencing remoteness into GIS

Current methods of estimating off-road accessibility rely heavily on guess-work, local knowledge and the manual application of Naismith's 1892 Rule. Although several modifications have been applied to this general rule of hill walking over the years, the basic rule and its method of application remain the same after over 100 years.

At a local level it is possible to develop a model that takes into account a range of topographic variables that influence on-foot travel times. This can be achieved by incorporating Naismith's Rule and more recent corrections (see section 12.3.3) into a GIS model.

Most GIS packages now provide a whole range of generic spatial modelling tools with which to build customised solutions to particular problems. Macro languages also allow the integration

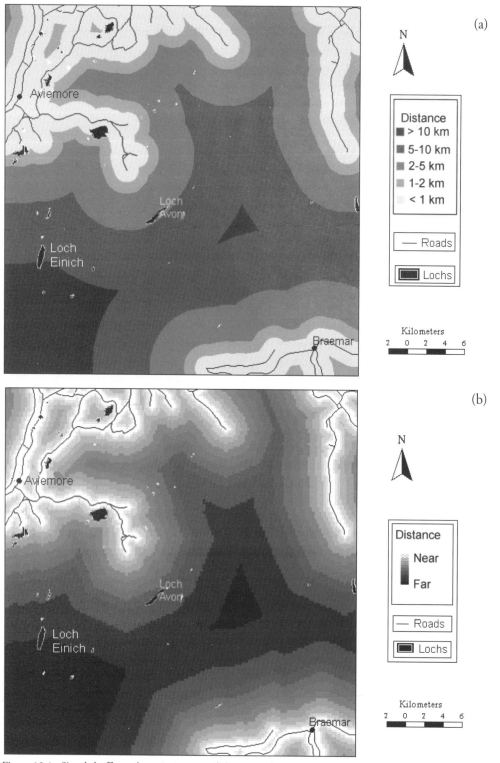

Figure 12.1. Simple buffer and proximity accessibility maps for the central Cairngorms. (a) Buffer zones around local roads at 1, 2, 5, and 10 km distance. (b) Proximity surface for local roads.

of external models into the GIS modelling environment. While to the best of the authors' knowledge no specific off-road pedestrian access models of the kind described here exist within any GIS package, it is clearly possible to build one using the GIS as a spatial modelling toolkit.

12.3.3 Implementing Naismith's Rule within a shortest path algorithm

W.W. Naismith was a founder member of the Scottish Mountaineering Club and a keen walker. His basic rule is still used to obtain a rough estimate of the time required for a given walk (Aitken, 1977; Langmuir, 1984). The basic rule states that a walker can maintain a speed of 5 km/h on level ground, but half an hour needs to be added for every 300 m of ascent. Several refinements have been made to Naismith's Rule. These range from Tranter's Correction, that takes an individual's fitness level and fatigue into account, to simple corrections that assume Naismith to be an optimist and so add 50% (Langmuir, 1984). Aitken (1977) made refinements according to ground conditions. This assumes that 5 km/h can be maintained on paths, tracks and roads, but is reduced to 4 km/h on all other terrain. Langmuir (1984) made the following further refinements: Naismith's Rule of 5 km/h plus 0.5 hour per 300 m of ascent, minus 10 minutes per 300 m descent for slopes between 5° and 12°, plus 10 minutes per 300 m descent for slopes greater than 12°. It is thought that the rule is generally applicable for reasonably fit hillwalkers negotiating typical terrain under typical weather conditions. However, further corrections can be made to allow for variations in terrain and conditions under foot, prevailing weather, steep ascents/descents, fitness and load carried.

Using Naismith's Rule it is possible to calculate the time taken to traverse a set of cells in a digital elevation model (DEM) by taking gradient and slope direction relative to direction of travel into account. A DEM is defined here as a digital model of height (elevation or altitude) represented as a regularly spaced grid of point height values. Values of slope (gradient) and slope direction (aspect) can be calculated from the DEM. Accessibility from different directions relative to the same point in the landscape should be considered and the shortest path or access time taken into account. Using this approach it is possible to design a model that calculates the time taken to walk from single or multiple origin points to any destination on the terrain surface. Due to the fact that it is unknown which route a walker will take, the model only considers the quickest possible path.

The model described here integrates Naismith's Rule with Dijkstra's shortest path algorithm (Aho *et al.*, 1974). Dijkstra's algorithm works by considering the relative costs of moving through each of the cells in a matrix. Costs are represented by impedance values in the cell matrix. In order to implement Naismith's Rule within Dijkstra's algorithm four different cost matrices were used. These include a heights matrix, a distance matrix, a trace matrix, which marks all the cells that have been dealt with, and a results matrix, the values in which are changed during the analysis process. This process has been automated within the Arc/Info GRID module and custom C code. For a detailed description of the implementation of the algorithm see Fritz and Carver (2000).

Using this approach it is possible to define remoteness surfaces for any landscape. Figure 12.2 shows a DEM of the Cairngorm overlaid with roads, rivers and lochs. Figure 12.3 shows the resulting remoteness surface from Naismith's Rule in Dijkstra's algorithm as applied to the DEM in Figure 12.2 for a single access origin at the Day Lodge car park near the Cairngorm ski area.

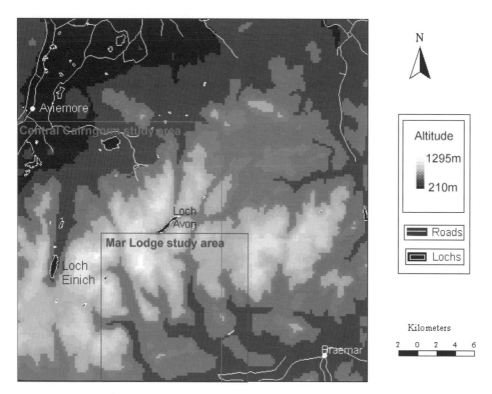

Figure 12.2. Central Cairngorms DEM with road network and water features.

12.3.4 Refinements to the algorithm

The most obvious improvement to the basic algorithm is the ability to handle more than one origin point and linear origin features such as roads. Remoteness from all roads surrounding the Cairngorm is shown in Figure 12.4. Other improvements can be made to the basic Naismith's/Dijkstra's algorithm described above to make the results more realistic in relation to the earlier definitions of remoteness. Some of the geographical factors affecting off-road access times can be incorporated into the algorithm by modifying the distance matrix according to additional cost/push factors. These include ground cover, underfoot conditions, presence of tracks and footpaths, prevailing weather conditions and the effects of barrier features such as lochs, crags, unfordable rivers and private land. Figure 12.5 shows the effects of linear access features (roads, tracks and footpaths) with the addition of barrier features (lochs and crags) and underfoot condition data. Here, underfoot conditions were classified as 'easy going' and 'hard going' for demonstration purposes while Naismith's Rule was modified to show walking speeds of 5 km/h on paths and tracks and 4 km/h in all other areas as suggested by Aitken (1977).

Comparing the distance maps in Figure 12.1 with the remoteness maps in Figures 12.4 and 12.5 shows how terrain and other geographical features can start to make a significant difference to mapped off-road access times. It is clear that the greater the relief and the more barrier features that are present, then the greater the divergence between simple linear distance maps and the remoteness mapping techniques developed here. The degree of difference between the maps in Figure 12.1 and those in Figures 12.4 and 12.5 can be

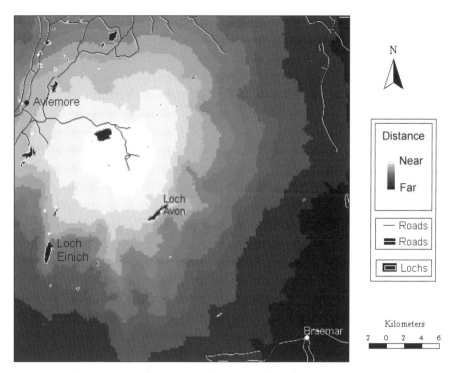

Figure 12.3. Remoteness surface for central Cairngorms based on single access origin (National Grid Reference 298900, 806300)

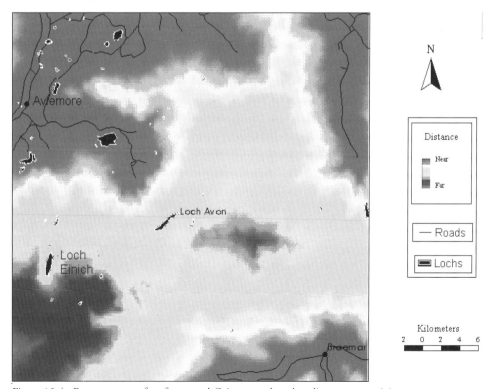

Figure 12.4. Remoteness surface for central Cairngorms based on linear access origin.

represented as residuals, where residuals are calculated as the difference between the two maps by simple arithmetic subtraction. These are shown in Figure 12.6.

12.4 Potential applications of remoteness mapping

It is suggested that the remoteness mapping technique described above may have several applications within the field of landscape management, including landscape character mapping, wild land mapping, planning, modelling carrying capacities, trip planning, and search and rescue operations. Planning, wild land mapping and zoning of protected areas are each considered further.

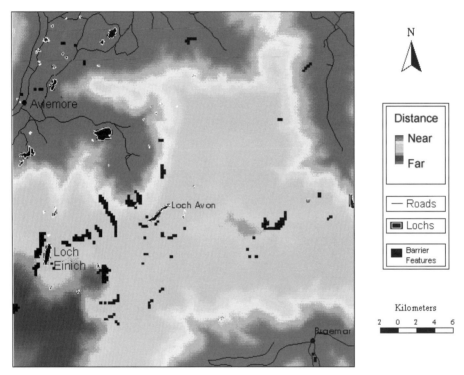

Figure 12.5. Remoteness surface for central Cairngorms based on linear access origin including footpaths, barrier features and land cover information.

12.4.1 Planning

Remoteness mapping may have several applications within the planning field. In particular, remoteness maps may be useful in route planning for new or upgraded footpaths, tracks or roads, evaluating new hydroelectric schemes and the effect of associated access roads, and assessing the impacts of new recreational developments. In this context, information on remoteness may be useful in helping consider planning applications for new developments. Using the techniques described above it would be possible for any proposed development to create 'before' and 'after' remoteness maps as part of the associated environmental assessment.

The proposed funicular at Cairngorm is an interesting issue affecting the Cairngorm area that may have a considerable impact on access to the Cairngorm plateau and on its remoteness. Current plans suggest that the development will actually reduce access for

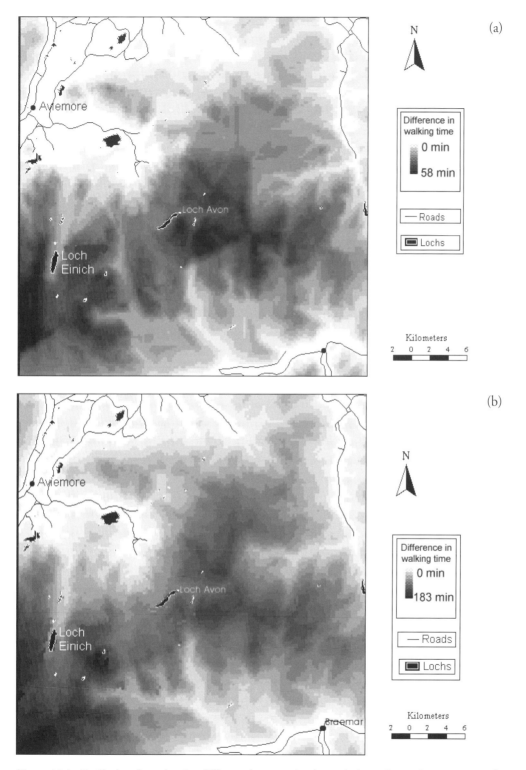

Figure 12.6. Residual surfaces showing difference between simple proximity surface and remoteness surfaces. (a) Comparison of Figure 12.1(b) and Figure 12.4. (b) Comparison of Figure 12.1(b) and Figure 12.5.

walkers and mountaineers to the plateau by restricting free car parking at the foot of the funicular only to funicular users and charging high parking fees for non-funicular users. This means walkers and mountaineers will be more likely to park further down the Allt Mór valley towards Loch Morlich, thereby increasing the time required to reach the plateau or other destinations such as the Lairig Ghru. Conversely, the increased number of people reaching the plateau via the funicular and the expansion of the Ptarmigan restaurant at its top will not increase the accessibility of the plateau itself since funicular users will not be allowed out onto the plateau from this point. This is illustrated in the two maps shown in Figure 12.7 that show remoteness before and after funicular development. The presence of the redeveloped Ptarmigan restaurant will, however, reduce the perceived remoteness of the plateau by its mere presence. This is not shown in Figure 12.7, but could be included in further analyses showing the effects of intervisibility on perceived remoteness.

12.4.2 Wild land mapping

There is a great deal of popular interest in the idea of wild land within the Scottish Highlands and other areas of the British Isles. This is amply demonstrated by the recent rash of glossy coffee table picture books with the words 'wilderness' or 'wild' in the title (e.g. Bellamy and Gifford, 1990; Linklater, 1993; McNeish and Else, 1997). Although these are nice to look at, a more scientific approach is needed to strengthen arguments for or against such an ideal. Recent work has explored the use of GIS and the wilderness continuum concept to examine the spatial distribution of wild land within Britain (Carver, 1996). The wilderness continuum concept states that for any area there is a range of landscape conditions from the wildest to least wild based on the degree of environmental modification by human action (Nash, 1982; Hendee *et al.*, 1990).

Recent work, principally in Australia and New Zealand, has used GIS modelling techniques to map this continuum for land management purposes (Lesslie and Taylor, 1985; Lesslie and Maslen, 1995). Generally there are four broad geographical factors that combine to define wilderness values. These are

- remoteness from settlement;
- remoteness from mechanised access;
- apparent naturalness (lack of human structures in the landscape); and
- biophysical naturalness (naturalness of the flora and fauna).

This work has been criticised on the basis that it over simplifies the complex inter-relationships between landscape character and ecology that define wilderness, largely through its use of ordinal data. In particular, the Australian National Wilderness Inventory (NWI) uses numerical scaling and combination techniques (map overlay) that may be construed to be inappropriate for use with ordinal data (Bradbury, 1996).

The remoteness mapping techniques described here have the benefit of being ratio data and therefore immune from this type of criticism. When combined with other ratio data sets describing environmental quality, the remoteness maps described here can be used to modify and improve the wilderness mapping techniques of the NWI. Work has adapted the NWI methods by replacing the ordinal overlay techniques with multi-criteria evaluation (MCE) methods (Carver, 1996; Fritz and Carver, 2000).

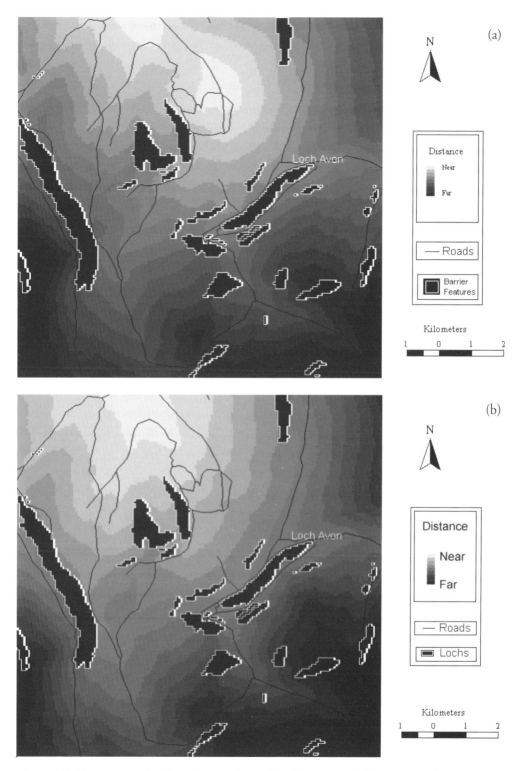

Figure 12.7. Remoteness surface for Cairngorm ski area based on nearest single point access feature. (a) Before the funicular development. (b) After the funicular development.

If the more open ended approach to wilderness definition advocated by Nash (1982) is adopted, then a GIS-based MCE approach to mapping the wilderness continuum is much more appropriate than other more deterministic methods commonly used in GIS-based analyses. This is because, like the continuum concept itself, MCE methods are not restricted by the necessity to specify rigid thresholds or criteria in defining where an entity like wilderness begins and ends. To meet a particular objective (in this case the mapping of wilderness quality) it is often necessary to consider and evaluate several criteria. This can be achieved using MCE, the basic aim of which is to evaluate a large number of geographical locations in the light of multiple and possibly conflicting measures of wilderness value. In doing so it is possible to generate rankings of the different locations according to their overall wilderness quality.

The application of MCE allows the mapping and combination of relevant datasets without the data loss associated with ordinal techniques (Janssen and Rietveld, 1990; Carver, 1991). The MCE approach also allows the specification of preference scores or weights for each of the data sets in such a way that personal perceptions as to the relative importance of factors can be included in the analysis. This makes the technique particularly suitable for mapping the wilderness continuum, especially in the light of Nash's (1982) comment about the differences in wilderness perception held by people of varying levels of wilderness experience.

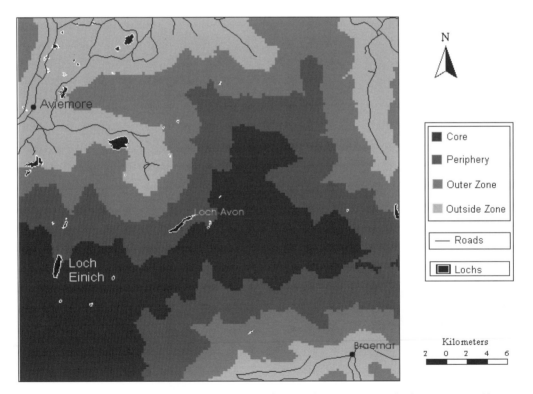

Figure 12.8. Core, periphery, outer and outside zones for central Cairngorm area land management (this is a reclassification of the remoteness surface in Figure 12.5 into four zones based on specified pedestrian travel times from the nearest road).

12.4.3 Zoning of protected areas

One further application area for remoteness mapping is in the zoning of protected areas for management and decision support purposes. With the increasing number of people participating in some form of outdoor recreation it is becoming more and more necessary to employ zoning techniques within Britain's national parks and other protected areas. Remoteness surfaces based on the geographical factors described here may be useful in assisting park managers to decide on the exact boundaries of these zones and their subsequent management.

Present proposals for National Parks in Scotland add an interesting and very applied dimension to the current work. Figure 12.8 demonstrates how the access surfaces shown in previous maps may be reclassified into different categories to map what may be termed 'core', 'periphery' and 'outside' zones within the Cairngorm case study area. The map shows a direct comparison to those zones generated by reclassifying the simple proximity map from Figure 12.1(b) with those generated by the more realistic remoteness surface from Figure 12.5. While it is not suggested that these zones fully reflect the situation in the Cairngorm, they do form a useful demonstration of the technique.

The main difficulty with this particular technique is in the definition of the threshold remoteness values that are needed to classify the remoteness map into discrete zones. This choice is probably best based on local knowledge and experience and the results combined with other information on wildlife habitats, biogeographical zones and land use.

12.5 Conclusions

Whereas remoteness may be difficult to define, it is possible to develop GIS-based models that map remoteness as a time/access surface. The coupling of the widely accepted and modified Naismith's Rule with a tried and tested shortest path algorithm allows off-road access times to be calculated for any roadless area. The resulting remoteness surfaces can be used for a variety of applications from landscape character studies to zoning of protected areas.

While the remoteness mapping techniques have had the effects of barrier features, footpaths, tracks and ground conditions, added to the model, further work needs to be done on both empirical testing and adding more walker-centred information. This information includes the effects of prevailing weather conditions and seasonal variations in climatic factors on walking times and the presence of human structures or impacts in the landscape on perceptions of remoteness.

References

Aho, A.V., Hopcroft, J.E. and Ullmann, J.D. (1974). *The Design and Analysis of Computer Algorithms*. Reading, Addison-Wesley.

Aitken, R. (1977). *Wilderness areas in Scotland*. Unpublished PhD thesis, University of Aberdeen.

Bellamy, D. and Gifford, J. (1990). *Wilderness Britain? A Greenprint for the Future*. Oxford, Oxford Illustrated Press.

Bradbury, R. (1996). Tracking progress: linking environment and economy through indicators and accounting systems. *Proceedings of Australian Academy of Science Fenner Conference on the Environment. Sydney, Institute of Environmental Studies. University of New South Wales, 1-7.*

Burrough, P.A. (1983). *Principles of Geographical Information Systems for Natural Resources Assessment.* Oxford, Oxford University Press.

Carver, S. (1991). Integrating multi-criteria evaluation with GIS. *International Journal of Geographical Information Systems,* **5**, 321-339.

Carver, S. (1996). Mapping the wilderness continuum using raster GIS. In Morain, S. and López Baros, S. (Eds.) *Raster Imagery in Geographic Information Systems.* New Mexico, Onword Press, 283-288.

Cuddy, S.M., Davis, J.R. and Whigham, P.A. (1996). Integrating time and space in an environmental model to predict damage from army training exercises. In Goodchild, M.F., Steyaert, L.T., Parks, B.O., Johnston, C., Maidment, D., Crane, M. and Glendinning, S. (Eds.) *GIS and Environmental Modelling: Progress and Research Issues.* Colorado, GIS World Books, 299-304.

Douglas, D.H. (1993). Least cost path in GIS. *University of Ottawa, Department of Geography Research Note* 61.

Ewert, A.W. and Hollenhorst, S.J. (1997). Outdoor recreation and its implications for wilderness. *International Journal of Wilderness,* **3**, 21-26.

Fenton, J. (1996). Wild land or wilderness - is there a difference? *ECOS,* **17**, 12-18.

Fritz, S. and Carver, S. (2000). *Accessibility as an Important Wilderness Indicator: Modelling Naismith's Rule.* Leeds, School of Geography, University of Leeds.

Hendee, C.J., Stankey, G.H. and Lucas, R.C. (1990). *Wilderness Management.* Fort Collins, Colorado, Fulcrum Publishing.

Janssen, R. and Rietveld, P. (1990). Multi-criteria Analysis and GIS: an application to agricultural land use in the Netherlands. In Scholten, H.J. and Stillwell, J.C.H. (Eds) *Geographical Information Systems for Urban and Regional Planning.* Amsterdam, Kluwer Academic Publishers, 129-139.

Langmuir, E. (1984). *Mountaincraft and leadership.* The Scottish Sports Council/MLTB. Leicester, Cordee.

Leopold, A. (1921). The wilderness and its place in forest recreation policy. *Journal of Forestry,* **19**, 718-721.

Lesslie, R. and Maslen, M. (1995). *National Wilderness Inventory Handbook of Procedures, Content and Usage, 2nd edn.* Canberra, Commonwealth Government Printer.

Lesslie, R.G. and Taylor, S.G. (1985). The wilderness continuum concept and its implication for Australian Wilderness Preservation Policy. *Biological Conservation,* **32**, 309-333.

Linklater, M. (1993). *Highland Wilderness.* London, Constable and Company.

McNeish, C. and Else, R. (1997). *Wilderness Walks.* London, BBC Books.

Nash, R. (1982). *Wilderness and the American Mind.* New Haven, Yale University Press.

Oelschlaeger, M. (1991). *The Idea of Wilderness.* New Haven, Yale University Press.

Outdoor Recreation Resources Review Committee (1962). *Wilderness and Recreation - a report on resources, values, and problems. Study Report 3.* Washington DC, US Government Printing Office.

Sedgewick, R. (1984). *Algorithms.* Reading, Addison-Wesley.

13 Approach to Landscape Character Using Satellite Imagery and Spatial Analysis Tools

Hubert Gulinck, Hans Dufourmont and Ingrid Stas

Summary

1. Landscape character is insufficiently used as a concept outside Britain.
2. An appropriate landscape character methodology would greatly improve planning strategies for peri-urban areas.
3. Land cover data derived from satellite imagery contain valuable information about features that strongly affect the landscape character of rural areas.
4. Radial analysis of open space in a land cover data grid yields numerous indices of landscape character in an almost continuous way.
5. Spatial indices can be combined in order to create preliminary landscape character models.
6. In the semi-urban study area, people reacted consistently against visual urbanisation, which can be mapped accurately using the given datasets and spatial analysis techniques

13.1 Introduction

The fact that the expression 'landscape character' can be an operational flag for country survey programs, scientific conferences and countryside policies should be appreciated. One of the obvious points of the British landscape character concept is the coverage of the entire territory and the attention to all landscapes, not just pre-selected settings of heritage, amenity or ecological value. Although a similar expression may exist in other languages, the operational strength is generally weaker. In the Dutch language, for instance, *landschapskarakter* is a rather idiosyncratic word used by landscape architects. It lacks semantic power to attract the necessary conditions for a systematic and functional modelling of the territory. In general, the fuzziness and fragmentation of landscape semantics may partly explain the fragmentation of policies and research programmes, and also undesirable changes in the landscape itself.

Nevertheless, in Belgium and elsewhere in Europe there is a strong awareness of threatened landscape character or identity, but landscape policies often lack comprehensiveness, and concern partial approaches to landscape character. A few examples can clarify this.

- The regional diversity of landscape characteristics is promoted through building permits in rural settlement areas in Wallonia (southern Belgium) (Ministère de la Région

Wallonne, 1996). The traditional map of physiographic regions, based on geology, soils and landform, serves as a landscape character reference.

- In Flanders (northern Belgium) the map of traditional landscapes was used as a landscape character reference in the 1996 Structure Plan (Vlaamse Regering, 1996). Traditional landscapes, however, refer to historical rather than actual characteristics. A refined and systematic survey of historical characteristics is being executed. Data concerning biotic landscape values are being collected in a variety of programmes and provide a strong focus on nature.

- In the same Structure Plan, the notions of urban versus rural areas, open spaces and landscapes have gained political and strategic strength in the demarcation of urban areas, protection of rural areas and the environment in general, but a typological framework for describing modern landscapes is missing.

Flanders, in particular, can be considered as a good laboratory for developing and testing innovative landscape character approaches, taking into consideration the high degree of urbanisation, the strong fragmentation by infrastructure and in general the very dynamic land use. In the period between 1985 and 1995 the number of parcels of land with a registered construction element increased by over 13 per cent. The number of landscapes cut by major infrastructure lines reached 44 per km^2 cell on average. The road traveller through Flanders has in many, if not most, areas a strong impression of landscape cluttered by ribbon development and dispersed constructions. In contrast, Wallonia has retained much more of the rural character of a mid-European upland area.

In Flanders, both historic and modern land use dynamics have led to a serious compartmentalisation. Since 1950 construction has become a key visual character. The boom in urbanisation has strongly decreased the regional diversity of landscapes. In a few decades, traditional agricultural landscapes have lost many of the specific characteristics of field patterns, field margin vegetation, crop types and rural buildings.

The need for an appropriate landscape classification is expressed in all modern planning concepts, such as in the Spatial Layout for Benelux. In relation to this, the concept of 'neo-rural' areas was introduced (Gulinck and Dortmans, 1996). Innovative and systematic 'landscape character' programs could substantially enhance the position of landscape in the different planning policies.

Brooke (1994) described the methodology for landscape character assessment of the Countryside Character Programme; it used a unique combination of critical judgement, electronic analysis and public perception. This chapter supports Brooke's approach, but starts from information provided through land cover data and spatial analysis tools. The results of this analysis are considered as spatial hypotheses concerning landscape character and value. They are not an objective in their own right; rather they are meant to help catalyse the dialogue between spatial data providers, landscape and spatial planners, and the general public.

13.2 Methodology and example

This chapter illustrates the use of land cover data, derived from satellite imagery, in combination with external GIS data, as input for a preparative outline of landscape character. The pilot status of such a project should be stressed. The input sources listed below certainly do not give access to many fine grain elements, nor to subjective and cultural interpretations of the living landscape. On the other hand, such input sources are increasingly available. There are four methodological steps towards landscape character definition.

13.2.1 Land cover data sets

In a strict sense, land cover is the biophysical state of the earth's surface (Turner *et al.*, 1995; Scott *et al.*, 1993). This definition is fashionable in eco-environmental research with emphasis on such topics as carbon cycling and geomorphology. Some experts emphasise vegetation and botanical characteristics. However, in most instances land cover is understood as the set of classes such as forest, urban land and open water. Such a definition is close to the culturally defined general mapping categories and to the definition of 'land use'. In strict terms, 'land use' refers to the manner in which land cover is manipulated and the intent underlying that manipulation (Turner *et al.*, 1995).

Remote sensing is designed to analyse the biophysical state of the earth. This biophysical state is expressed in spectral signatures, which in turn are the key to land cover mapping (in its categorical sense), the most popular application of terrestrial remote sensing. Because of the implicit link to externally visual land surface characteristics, mapping traditions and land use, land cover datasets derived from remote sensing are biased towards the notion of landscape character.

For the 13,000 km² of Flanders, 1995 Landsat data were subjected to a basic classification procedure into the elementary six classes: built surface, water, forest, permanent grassland, bare soil, and agricultural crops. Classification upgrading was done by multitemporal combinations, texture analysis, filter techniques, the application of vegetation and other indices, location rules, and the linkage to external data layers such as soil maps or digital relief models. The land cover list could then be expanded to over 15 standard classes. With a careful classification design over 85 per cent accuracy is obtained using good quality images of appropriate calendar date. The limited success of classifying linear landscape elements such as sunken roads, hedgerows and small streams, even using panchromatic SPOT data, is certainly a weak point, especially in the framework of landscape character analysis.

Each resolution level yields its types of mixed pixels for a specific geographical area. Mixed pixels are generally a nightmare for land cover classifiers, but can sometimes be profitable in a discussion about landscape character. The mixed category of loose construction is a particularly interesting one. It is spatially composed of contrasting units with horizontal dimensions smaller than the 30 m resolution limit for LandsatTM. In the land cover database for Flanders, it contrasts with densely built units and corresponds to fine grain residential complexes with a minimum of 30 per cent physical built cover. This category has to be considered as one of the major modern landscape types, especially associated with incoherent peri-urban areas and with ribbon development.

These land cover data layers can be combined with external datasets such as the Belgian digital elevation model, the road network, a soil association map, etc. These contribute greatly to basic information concerning landscape character.

With the introduction of higher resolution digital remote sensing images (down to pixels of 1 m square) in the near future, a substantial refinement in land cover datasets is to be expected. Such input data will be especially valuable for the identification of cover types in highly fragmented or complex landscapes. On the other hand, the extraction of information from high-resolution remote sensing data will ask for much more complex procedures, which will not be discussed here.

Intrinsically, landscape character refers to the integration of information over some 'viewshed', or a perceived landscape unit, that generally has a size much larger than the 10–30 m medium resolution of SPOT or Landsat. Therefore coarse resolution data are not necessarily less useful for landscape character assessment. In the Flemish regional environmental monitoring programme, 20 m land cover grids were aggregated to 1 km grids by two techniques: the area proportion of the cover categories, and counting the unconnected objects belonging to the different categories (Gulinck *et al.*, 1998). This can be related to criteria of landscape complexity and diversity.

The CORINE land cover database for Europe (European Commission, 1993) was based on visual interpretation of Landsat imagery with 25 ha as the minimal significant area unit. It was designed as a land cover database, but its legends show land use references as well. Some of its categories are landscape rather than cover units, as aggregates of use and cover types, or undifferentiated fine grain complexes. Despite its relatively coarse 25 ha spatial resolution, this database is a valuable input for initiating a landscape character framework at European level. Finally, even coarse resolution input remote sensing data such as 1km VEGETATION (Centre National d'Études Spatiales & Joint Research Centre; http://www.spotimage.fi/date/images/vege/vegetat/htm) may yield valuable information for landscape character assessment as the information content of the pixels is determined by the dominant land cover characteristics.

13.2.2 Spatial indices of landscape character

Land cover approaches emphasise the elementary image units (pixels) and therefore yield little information about landscape character. In applying spatial metrics to land cover datasets, substantial complementary information about landscape character can be derived. Different types of landscape metrics exist in commercially available software packages. One of the best known worldwide is Fragstats (McGarigal and Marks, 1995); this calculates landscape indices of area, shape, edge, number, distance and similarity of patches, land cover classes or whole landscapes. Although certainly useful on the way to develop landscape character models, a different set of metrics was used for this chapter.

The core idea is to base morphological measurements on a 'radial mode' (Dufourmont *et al.*, 1991; Gulinck *et al.*, 1993). Each pixel of open space (this is outside built and wooded landscape patches) is considered as a potential observation point, from which the surrounding landscape is measured around 360°. The radii in a disk centred on each pixel were regularly sampled in a number of steps to yield detail about the morphological structure of the landscape. Distances to the 'radial obstacles', the nature of these obstacles and the type of ground cover in the radial lines are the parameters for defining and distinguishing landscape morphological types (Plate 19). Several such indices are discussed below.

The emphasis on 'radial' features makes it clear that this analysis is comparable but not necessarily equal to an analysis based on 'visual' features. For instance, in a certain context one might wish to measure the average distance to urban features. In another context, the degree with which urban features are visually masked by vegetation may be of interest. A third example might be the degree of 'visual greenness' of a vantage point.

A single radial run from a point of a land cover dataset yields a variety of morphological indicators of landscape character. The following list is not exhaustive.

- Average radial length (ARL): this is simply calculated as the mean of the distance from the observation point to the visual (or other) boundary. A maximum visual length of 1400 m was assumed as a truncation value for longer views. The ARL is an excellent index of size or scale of open space. In a region affected by peri-urbanisation, rural spaces tend to be fragmented and to decrease in size over time and with closer distance to the urban centres.
- Radial complexity: this index gives an idea of the structural complexity of the landscape, typical for a landscape affected by dispersed construction or by other factors such as the historical fragmentation of woodland (Plate 20). It is calculated by normalising the summed differences in consecutive visual radii by the mean, or simply as the standard deviation of the radii.
- Diversity of surrounding land cover: this index measures the number of different types of land cover that are encountered within a given distance. In a variant, one can measure the sequence of land cover types as the view orientation turns around.
- Boundary characteristics: a series of indices can be defined for measuring the type and quantity of 'visual boundaries'. These indices are particularly important for measuring such impacts as urban encroachment or the degree with which views have a woody signature.

13.2.3 Morphological landscape types as a prelude to landscape character mapping

Using appropriate numerical tools, landscape types can be derived from combined datasets of land cover data, morphological data and collateral data. Using a limited set of spatial indices as described here, it is possible to derive preliminary landscape character maps, with emphasis on the morphology of land cover. Gulinck *et al.* (1993) discussed a standard typology procedure that proved to discriminate well between different cultural landscapes in an innovative way. In combination with a digital elevation model, a road database, etc., it allows one to assess the relative impacts of relief, urbanization, infrastructure and dispersion of woodland and trees on the visual characteristics of open space (or structural qualities in general), either separately or in combination.

13.2.4 Linkage to real world landscape character appraisal

The resulting classes can be used as a stratification basis for further specifying the landscape character through more classical assessment methods and direct field observation. As an example, a pilot study was done in a 20 x 15 km peri-urban area east of Brussels. Three factors are considered essential in the discrimination of landscape character type and in the visual appraisal of landscape; these are relief, tall vegetation (woodland patches, tree screens) and construction features (dispersed, linear and concentrated development). It is assumed that in a peri-urban area, rural character is generally preferred over the visual intrusion of urban features.

To test this assumption, 195 photographs taken in 1997 were selected from an irregular grid with a minimum 500 m interval. At each location there were four non-overlapping photographs (N, E, S, W). The photographs were subjectively sorted into seven classes determined by two factors: the degree of urbanization, and openness (distance to visual screens). Topography was provisionally omitted because of its limited expression in this region and because it is not a dimension of the input satellite imagery. Next, the photographs were sorted in five degrees of appreciation by a sample of 20 people in each of three categories, local inhabitants, external non professionals, and external landscape professionals. In a preliminary test no significant difference in preference by these groups could be found. Moreover, there appeared to be a strong degree of correspondence between public approval and the landscape typology (Figure 13.1). In this study area, the presence of construction is a factor that is not appreciated, whereas trees and woodland increased the appreciation. Although this is a preliminary landscape appraisal model, it demonstrates that, for this region, the important physical and structural characteristics that play a role in landscape perception are easily derivable from satellite imagery.

The next step will be to find out whether morphological types, as derived from land cover, can match the ranking of photographs into the seven classes. In a preliminary test, the

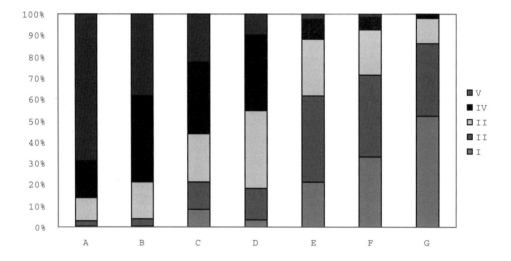

Figure 13.1. Classes of appreciation of photographs versus viewed landscape types. Landscape types are A (densely built), B (open and built), C (green and built), D (mixed), E (open and rural), F (half open and rural) and G (rural). 'Open' referes to visually open. The last three classes are essentially visually free from constructions. The Y-axis gives the percentages scoring in 5 classes of appreciation: I is highly appreciated and V is unappreciated, with the classes II to IV being intermediate between these two extremes.

morphological criteria were the kind of the visual barriers (green or construction) and the distance to these screens. Distances were set at 300 m, 600 m and 900 m in order to match the range of visual distances as observed in the field. From these criteria seven morphological classes were defined so as to match the types derived from the field photographs. Each photograph could be compared with the corresponding pixel. It was possible to match over 90 per cent of pixels classified by morphological type with the classes of field photographs.

13.3 Discussion and conclusions

Existing landscape classification concepts often lack direct references to modern land use and land cover. New methods are needed to analyse complex peri-urban areas and construct quantitative indices for evaluation and planning. Land cover and/or land use maps can be considered as objective information bases for expanding such classifications.

The methodology presented can be considered as an effort to give the abundantly available satellite imagery, and the derived or co-existing regional GIS-information, an original and meaningful application. There is a substantial archive of medium resolution satellite data over a time span of almost 15 years over most areas of Europe. In the future, there will be a suite of satellite data of different resolution, from the metre to the kilometre range, allowing us to capture landscape characteristics of different spatial graininess.

Because the satellite imagery is good at separating major structural and morphological characteristics such as urban development, woodland, water, open spaces, bare soil, etc., it can yield valuable indices about landscape character. In combination with digital elevation models and other data sets, fairly detailed and virtually continuous preliminary landscape character maps can be produced. The residual discontinuity is fine grained and determined by the grid resolution of the original satellite datasets. These maps can serve either as a stratification framework for field investigations towards landscape character definition, or more directly as spatial hypotheses about landscape character types. Effects of land cover changes can easily be quantified.

The radial technique provides metrics that, to a certain degree, allow quantification of the viewshed structural characteristics, and hence can be used to model the morphological character of any location. Substantial refinements to the technique are possible, such as through the analysis of multiple visual objects in the sight lines, especially in hilly areas. Improvements are also possible in the modelling of preference scores or conventional landscape character definitions that utilize the landscape structural parameters.

At this stage the methodology lacks a correspondence with landscape character maps that have been derived from more traditional approaches. A rather more mechanistic approach may be discerned as opposed to an acceptance of more intuitive and cultural factors in landscape character studies. On the other hand, the method presented in section 13.2.2 supports rapid prototypes of landscape character over large areas in a single computer run. This in itself may provide impetus for promoting the concept of landscape character in areas where there is low academic or strategic interest. At the European level, the general application of satellite imagery may help to develop continental coverage of landscape character assessment on a comparable basis.

Acknowledgements

This paper has drawn from research financed by the Ministry of Science Policy (Brussels), the Flemish Land Agency and the Flemish Environment Agency.

References

Brooke, D. (1994). A countryside character programme. *Landscape Research*, **19**, 128-132.

Dufourmont, H., Gulinck, H. and Wouters, P. (1991). Relief dependent landscape typology derived from SPOT data. *Proceedings of the 2nd European Conference on Geographical Information Systems, Brussels*, 286-293.

European Commission. (1993). *CORINE Land Cover Technical Guide.* Luxembourg, European Commission.

Gulinck, H. and Dortmans, C. (1996). Neo-rurality. Benelux as workshop for new ideas about threatened rural areas. *Built Environment*, **23**, 37-46.

Gulinck, H., Walpot, O. and Janssens, P. (1993). Landscape structural analysis of central Belgium using SPOT data. In Haynes-Young, R.H., Green, D.R. and Cousins, S.H. (Eds.) *Landscape Ecology and GIS.* London, Taylor & Francis, 129-139.

Gulinck, H., Andries, A., Dufourmont, H., Dessers, E., Antrop, M., Martens, I. and Wiedeman, T. (1998). Fragmentation. In Verbruggen, A. (Ed.) *Report on the Environment and Nature in Flanders* 1996. Leuven, Vlaamse Milieumaatschappij, 361-373.

McGarigal, K. and Marks, B. (1995). *Fragstats: spatial pattern analysis program for quantifying landscape structure.* Pacific Northwest Research Station, USDA Forest Service.

Ministère de la Région Wallonne (1996). *Règlement Général sur les Bâtisses en Site Rural.* Jambes, Ministère de la Région Wallonne.

Scott, J.M., Davis, F., Csuti, B., Noss, R., Butterfield, B., Groves, C., Anderson, H., Caicco, S., D'Erchia, F., Edwards, T., Ulliman, J. and Wright, G. (1993). Gap analysis: a geographic approach to protection of biological diversity. *Wildlife Monographs, 123, 1-41.*

Turner, B.L., Skole, D., Sanderson, S., Fischer, G., Fresco, L. and Leemans, R. (1995). *Land-Use and Land-Cover Change Science/Research Plan.* International Geosphere–Biosphere Programme Report No. 35.

Vlaamse Regering (1996). *Structuurplan Vlaanderen.* Brussels, Ministerie van de Vlaamse Gemeenschap.

14 HIGH RESOLUTION GEOGRAPHIC IMAGE DATA FOR LANDSCAPE VISUALIZATION

Alun C. Jones

Summary

1.   The quality of the image base for landscape visualization is critical to the realism of the model.   The key to any visualization database is not in the data capture mechanism, the sensor and or satellite platform, but the data delivery system, the data retrieval system and in the geographical accuracy of the image data.

2.   The image data need more careful study and selection, as it is these image data from which we make our maps, extract landscape content and ultimately use in to make our decisions.

3.   Fitness for purpose is the key factor when determining the type and quality of geographic data required for any landscape visualization model.   When requiring an image to be made geographically accurate, the imagery can only be made to the same level of accuracy as the control data supplied.

4.   Availability of geographic data via the Internet will increase the uptake of such data and will eventually lead to a decrease in cost of data.   In the near future cost will not remain a central stumbling block to projects that require geographic data.

14.1 Introduction

The visualization of a landscape, particularly in 3D, is becoming recognised as an effective and highly informative means of qualitatively analysing environmental change (Loh *et al.,* 1992; Faust, 1995; Rowe, 1997).   However, as with all data modelling it is the quality of the base data, in this case either the terrain model or the image base, that has the most profound effect with regards to the success or failure of the visualization technique. Currently there exists, in the market, highly sophisticated software for generating terrain models from stereo images and draping them to present a landscape in 2.5D (Rowe, 1997). Imagery of sufficient quality and price is currently available via digital ortho-photography suppliers, but such suppliers have only localised coverage as opposed to global coverage.

With the new commercial satellites due to be launched over the next few years the possibility of building a high resolution (<1 m pixel) image and topographic database for landscape visualization, for any region in the world, will become a reality.   Yet these databases may be improved by the integration of multi-resolution databases, namely digital ortho-photography and additional high-resolution satellite imagery.

The key, however, to the value of such a database is not in the data capture mechanism, the sensor and satellite platform, but the data delivery system, the data retrieval system and in the geographical accuracy of the image data.   Only by producing a database that can deliver highly accurate and timely data in a form that all levels of user can use within any visualization system, will the database become a truly valuable tool in the decision making processes.

This chapter looks at what data are available for use within landscape visualization applications, the issues associated with obtaining these data, managing them and making such data available to all potential users. The chapter also examines the options for distributing such databases through online systems via Intranets and the Internet.

14.2 The data problem

In recent years there have been great leaps in the technology available to generate 3D simulations of environments for visualization purposes (Fisher *et al.*, 1993; M.J. McCullagh, pers. comm.). Once the domain of programmers and the UNIX or mainframe operating platforms, standard desktop computers now have the computing power to deliver, for a fraction of the cost, but at an acceptable level of speed, highly visual renderings of landscapes both in 2D and in 3D. However, we have yet to see such a concomitant rise in the quality and accuracy of the data required for such models. The result is confusion and frustration in landscape modelling; investigators often wish to see a greater level of detail within a model and also wish to interact with the model to a degree which is not possible due to the limitations of existing datasets.

The development of landscape visualization techniques, and the software environment with which to create realistic or rendered landscapes and convey those to a wider audience, does not need developing further. The area that needs the greatest attention is the data for the models and importantly the resolution, accuracy and availability of these data (Plate 21).

In terms of data for landscape visualization there are perhaps three key data types which make up the main content of a model. First level data (or terrain data) are critical to give the lie of the land and to enable the software to provide perspective and depth to the visualization. The terrain data, whilst critical to the overall model, do not necessarily register with the user as being a primary source of information. However, without such data the landscape is indeed flat and therefore the model fails to convey the realism of that landscape. Second level data (or image data) to be draped on the terrain model are, however, vitally important. It is these image data, either in the form of a cartographic image or a satellite or aerial photo, that give the context and content of the landscape and provide the most immediate data for decision making. The third level of data, and one that is only now becoming available due to advances in computer animation and technology, is that of objects. Objects can be placed on a model to provide added realism, in the form of tree stands, buildings, power pylons, etc. (see Plate 21). This last data form is often superficial and can only be added with some prior knowledge of the area or with some end goal in mind. These data are still in their infancy and it is this area that will see the greatest level of growth with advances in computer animation and simulation. They will lead to a truly accurate model of the real world on the desk top.

14.3 Fitness for purpose of geographic data

Determining at an early stage what is the prime purpose of the model and its intended result is key to building a relevant and workable model (Chrisman, 1991). As discussed above two key data types for landscape visualization in a 3D environment are the terrain data to give the elevation and perspective and the image data to be draped over this terrain to give the context information to the model. When determining what data are required, or when deciding to choose from what is available, the end goal of the model must be considered. This has a significant impact on the nature of the data used and, as with all data, it is vital that the optimum data type is used to avoid either over engineering the model or degrading the model to become unrepresentative of the real world.

The fitness for purpose of data can be classified in three ways, and with each a key criterion can be identified. Firstly, *measurement* is the accuracy and the ease with which features can be measured and these are primary factors. Secondly, *identification* in which both the clarity of the image and the scale of the image are important factors. Thirdly, *presentation* must be considered. If the end goal is simply to present a good image or model then colour and the seamless nature of the image are important.

A clearer view of the nature of the data required as an input to the model will be gained by taking into account these three factors. Table 14.1 shows the scales and resolutions available today for both image and terrain data. Although both types are important to the visualization models created, it could be argued that the features of the land and those represented by the image component of the data change more rapidly than does the underlying topography, represented by the terrain data. It is the image that needs more careful study and selection, as it is these image data from which maps are created and landscape content extracted for use in the analysis of the landscape. Due to both advances in technology and competition amongst the image data providers, users are now in a position to have, or soon will have, high quality image data for use in visualization environments. Such data will enable users to see individual trees, cars and buildings both from aircraft and from space. Data of this scale and quality bring the 'human' element to our decision making by enabling us to recognise, in the model, features that are readily identifiable.

Table 14.1. This lists some of the more common data types available on the market today. Note the inconsistencies of scale between terrain data and image data, and therefore care is required when both are merged together for landscape visualization.

Data type	Product name	Scale/resolution	Coverage
Terrain data	EDX	50 m grid spacing	UK, USA and other selected countries
	Ordnance Survey Landform Panorama	1:50,000 scale mapping, 10 m contour intervals	GB
	Ordnance Survey Profile	1:10,000 scale mapping, 5 m contour intervals	GB
	ETOPO30 Terrain data	1.1 km	Global
Image data	Landsat satellite	30 m	Global
	SPOT satellite	10–20 m	Global
	IRS-1C satellite	5 m	Global
	Aerial Photography	25 cm (approx. 1:1000 scale)	Selected areas
	Ordnance Survey Raster colour	1:50,000	GB
	Ordnance Survey Raster black and white	1:10,000	GB

14.4 Data Availability

1994 saw the signing of a Presidential order in the USA allowing commercial companies to use de-classified spy satellite technology for civil applications. Three US companies are competing to be the first provider of this technology. By late 1999 the first images from these satellite will be available on the market. Such satellite data will provide for the first time high-resolution data sufficient to see individual landscape elements and to view the change on a weekly, and even a daily, basis. Along with advances in the satellite data, the aerial photo providers also have new offerings which can improve the quality of landscape visualizations; for example Cities Revealed® has a 25 cm resolution aerial photo database (Jones, 1998). The advantage of aerial photography is that it is often easier to use, easier to purchase and provides higher resolution data than imagery from satellite systems. These developments enable the integration of both data types, one for clarity and resolution, the other for coverage and frequency of updates.

What this now presents is a conundrum of data richness. This in turn presents the potential user with a problem of selecting the right image for the job. To overcome that conundrum, a careful examination is needed of the end goals of the visualization and the fitness for purpose of the data for the visualization. One of the key factors in this is the resolution of the data selected for use in the model.

Data resolution can be defined in two ways. One definition is the true physical area covered on the ground by the smallest element of an image (termed a pixel). For example, the new satellites will be able to deliver 1 m pixel resolution data, and thus objects less than 1 m in area will not be detected. An exception exists here for linear features such as rails and road markings that are sufficiently distinct in either shape or colour from their surrounds that they stand out even though they are less than the resolution of the image. The other definition is one where resolution can be defined as the smallest object viewable in an image. For example, to detect a man hole cover that is 1 m in diameter may require four or five pixels of information to enable you to see what it is because a single pixel will just not give enough of a picture. The significant difference to remember here is that although many sensors offer identical image properties, i.e. 1 m square, what can be seen with that 1 m square can be significantly different. This difference depends upon sensor optics, film quality and flying height (if aerially based) and atmospheric conditions. The types of features that can be identified at particular resolutions are shown in Table 14.2.

Table 14.2. The table gives an approximation to the common features of the landscape that can be identified in comparison to resolution and scale of the image. by first determining what features need to be identified a better choice of the data scale and resolution can be made. The table does not account for magnification of the original image.

Feature to be identified	Approximate image scale and resolution to identify feature	
	Scale	Resolution
City areas	1:40,000	10 m
Buildings, roads	1:30,000	9 m
Trees, building condition	1:15-20,000	4-5 m
Tree location, height	1:10,000	2-3 m
Garden area, building access	1:5,000	1-2 m
Car type, pavement	<1:1,000	<1 m

Resolution is of fundamental importance to any project as it not only determines what can be seen, but also the provider of the imagery, its cost (Figure 14.1) and, something that is often neglected, the number of file volumes that are required to store the imagery. This has a secondary impact on the methods of distributing and integrating the data (P. Berger, pers. comm.; J. Kuiper, pers. comm.) into other systems and to other projects, particularly for Intranet and Internet delivery mechanisms.

For a doubling in resolution there is a fourfold increase in file size. Likewise there is an increase in cost, perhaps the most critical consideration of any project.

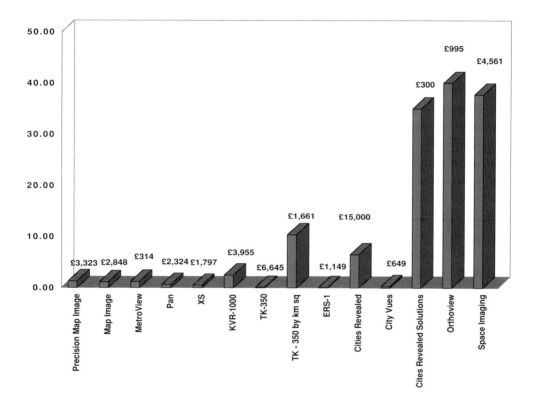

Figure 14.1. Comparative costs of satellite and aerial photographic data based on £ km^{-2}; minimum order required to purchase data at that £ km^{-2} rate is also shown above each data type.

14.5 Data management

With the advent of high-resolution data (0.25 m -1 m) user expectations as to what can be seen in an image are set to rise. Now, many aerial photography companies routinely delivery 25 cm or better resolutions. These data are sufficient to see people and even cracks in roads. As a result, user demands increase; however, what is often ignored is the data management that is required to enable high quality data to be displayed easily. For example, an area the size of Scotland (say 80,000 km^2), if imaged at 1 m resolution and supplied as common TIFF imagery in full colour, would yield a database of over 3.6 terrabytes, enough data to fill over 7,000 CDs. Now logic would dictate that such an area would be subdivided into more manageable subsets, but users could still be faced with databases in the 100 gigabyte region; managing such data volumes as recently as 1996 was daunting, if not impossible. However, with new technologies on the market, e.g. wavelet compression through MrSID™ – multi-resolution seamless image database, management of such file volumes is now a reality. Such 100 gigabyte files can now be compressed with minimal loss to only 5 gigabyte or less, and further splitting of these files can result in rapid delivery of these data over the Internet or Intranet.

The above serves to show how there has often been a trade off between the quality and the resolution of imagery with the complexities of managing and obtaining such data. Technologies are now available to overcome these barriers and very soon we shall not even talk about such complexities as they will be hidden from us in advanced interfaces, designed to deliver information – as opposed to today delivering just images, e.g. HTML.

14.6 Data access

'Plug and play' has been a key phrase for many software and hardware vendors. Until now it has not been associated with data. The concept really means that you simply plug in the device in question, or CD-ROM containing data that you wish to load, and after a few mouse clicks you access the required dataset. Data, and particularly geographic data, have often been the subject of a multitude of formats, delivery devices and projections. Particularly for image data, users have had to put up with endless arrays of CD-ROMs or other devices all containing a jigsaw of data that may eventually fit together. However, with new technologies for data display, compression and retrieval, a data management system delivering 'plug and play' is now a reality for data users (see Figure 14.2). When choosing

Figure 14.2. Data management interface to enable the viewing, user selection and distribution of large, multi gigabyte files. The software enables access to compressed data in MrSID format and uses a decompression tool to decompress data rapidly for use in other GIS visualization applications. Software such as this enables large volumes of data to be delivered rapidly over Intranets and Internets (image copyright The GeoInformation Group, 1999).

data, the data management and delivery must not be overlooked, as this will contribute to a significant portion of the project time; this is time better spent analysing the data.

Advances in Internet delivery mechanisms along with those designed for the more internal Intranets now mean that large image databases can be accessed through standard web browser technology. With developments in data management software and a growing awareness of the benefits of using Internet/Intranet solutions, access to large image databases from the desktop will soon become common place.

14.7 Data accuracy

Data accuracy is an issue sometimes overlooked and often over specified. Accuracy should not be confused with resolution; the former refers to the closeness of the geographic position of any feature in the image to its true geographic position whereas the latter is just the pixel size. As with most of the data issues addressed in this chapter, fitness for purpose is again the key factor. Consideration must be taken of the accuracy required in terms of measurements and the end decision required. Additional factors that must be considered are the quality of the data provided, and where imagery is used in a geographical framework (for use in a GIS), the accuracy of the control data. When requiring an image to be made geographically accurate the imagery can only have the same level of accuracy as the control data supplied. For example, if using a 25 cm resolution image (equivalent to 1:1,000 scale) and providing 1:25,000 scale mapping data for control, none of the resultant measurements will be accurate to 1:1,000 scale.

Also when a high degree of accuracy is required, for example between 0.5–1 m horizontal matching of the image data with its control data, there is an increase in cost. Again the end decision required from the model should be considered when specifying accuracy of the image and terrain data for the model. It is pointless to specify 10 cm accuracy when using 1 m imagery on a 50 m digital terrain model (DTM), even more so when the end result may only require a visualization without any measurement. By increasing the accuracy requirements there is an eventual rise in cost, both in terms of the data itself and in processing the data.

14.8 Data costs

It may now become apparent that when trying to specify data for a landscape visualization model there are numerous inter-related factors. Each has a varying degree of impact on the final model. However, these also impact greatly not only on the model quality itself but significantly on the cost associated with creating the model. The higher the specification for each of the components of the model, the higher will be the eventual costs. With that in mind, weighing up the value of the increased accuracy over the specifications of the final model must be undertaken so as to produce the optimum model for the project.

However easy and straightforward that statement may seem, in practice this is a complicated exercise. Figure 14.2 shows the variance in data costs for the more common image data that can be used in landscape visualization models. What is interesting to note is that although some data are extremely cost effective when viewed in terms of £ km^{-2}, a large amount of data often has to be purchased to get such low prices.

However, with advances being made in data transfer and data availability, data are becoming more cost effective and easier to purchase. There are three good examples of this effect.

First, 1999 will see the launch of the IKONOS satellite from Space Imaging. This will lead to imagery at 1 m resolution being made available in a commercial manner for any region of the globe (data at this resolution were, prior to Space Imaging, restricted to military uses). The image data and terrain data will also be available. This presents a new source for the building blocks of any landscape visualization model. In fact Space Imaging is already producing 'off the shelf' 3D visualization models complete with viewing software; this is currently only available for the US, but does show that pre-wrapped 3D visualization models will be available for other areas of the world in the near future. For the imagery, prices start at around \$54 km^{-2}, but at least 121 km^2 must be purchased. Although prices may seem a little high, within a few years prices should fall to enable users to access these high quality data within tight budgets.

Second, digital aerial photography giving 25 cm resolution imagery (approximately 1:1,000 map scale) is now available in single kilometre blocks. Data from Cities Revealed (www.crworld.co.uk) enable users access to extremely high quality data of small areas for use in visualization models. With data being as little as £10 km^{-2} this offers options for those projects where cost is a limiting factor

Third, the delivery of data and landscape models over the Internet, along with ordering and payment through the same channels, is fast becoming a reality. For example, Microsoft's Terraserver with its 1.1 terrabytes of high resolution Russian Spin-2 satellite data (over 2,000 images) is already enabling this. The key here is that once it is accepted by the mainstream section of the market it opens up a global purchase base and significantly brings the price down for these data. The concept of selling data at a low cost to many people is the only way that imagery will be made available at the prices we want to pay, i.e. less than £5 km^{-2}.

Cost will not remain a central stumbling block to projects that require geographic data. Global terrain data are now available for the price of your phone bill into a web site; image data are also available at the same cost. The stumbling block is to be able to obtain data of the right quality, for the right area, with the right accuracy and in a format that is ready to use. This is available for limited areas from aerial photo companies; however, for the whole globe it is yet to be a reality.

14.9 Conclusions

The 3D visualization often used to depict a landscape seems such a simple image, but it is not. If the end goal is to have a pretty picture then a simple model can often be created for the cost of the CD-ROM on which the data are provided. However, with this quality of data the model should only be utilised for just that purpose – a pretty picture. If, and it is here that the real future lies for 3D models, detailed analysis, placement of real world objects into the model and interaction between the user and the model is required, then high quality data are required. The saying 'rubbish in, rubbish out' is never more true than with geographic data for landscape visualizations.

References

Chrisman, N.R. 1991. The error component in spatial data. In Maguire, D.J., Goodchild, M.F. and Rhind, D.W. (eds.), *Geographical Information Systems, Principles and Applications, volume 1.* Harlow, Longman and New York, Wiley. pp. 165–174.

Faust, N.L. 1995. The virtual reality of GIS. *Environment and Planning B: Planning and Design*, **22**, 257-268.

Fisher, P., Dykes, J. and Wood, J. 1993. Map design and visualization. *Cartographic Journal*, **30**, 136-142.

Jones, A. 1998. How to build your own digital orthobase. *Mapping Awareness*, **12**, 46-49.

Loh, D.K., Holtfrerich, D.R., Choo, Y.K. and Power, J.M. 1992. Techniques for incorporating visualization in environmental assessment: an object-oriented perspective. *Landscape and Urban Planning*, **21**, 305-307.

Rowe, J. 1997. Is it real or data visualization? Windows-based packages bring terrain data to life. *GISWorld*, **10**, 46-52.

15 The Use of Information Technology in Landscape Planning

Peter Minto

Summary

1.   Although information technology (IT) provides an increasing array of tools suitable for landscape analysis, take up has been limited.
2.   Three recent systems for landscape characterization and evaluation are discussed, particularly in relation to the use made the techniques of IT.
3.   Some of the specific benefits provided by IT for landscape are discussed, and an argument made for the experimental use of these techniques by practitioners in the field.

15.1 Introduction

In discussing the use of information technology (IT) in landscape planning, it is tempting to remark that IT is not much used at all. With a few exceptions, IT is not used for landscape planning, or for evaluation, or even for categorisation. Conferences on landscape seem to be divided between those where people from universities and commercial bodies are developing new IT based techniques for analysing landscape, and the others where the potential end users, generally decision-makers and bureaucrats in government organisations, are still not using, or intending to use, these techniques. This chapter addresses why this might be so, and what might be done to remedy the problem.

To understand the present situation, it is always useful to look at the past. Early attempts to marry IT solutions with landscape issues were probably over ambitious and flawed because of a desire to employ numerical solutions. The general thrust was to evaluate by a process of quantification and measurement, something which seemed alien to the spirit of landscape appreciation. The wish to apply numbers to everything has a long history. "By weighing, we know what things are light, and what heavy. By measuring we know what things are big, and what short. The relation of all things may thus be determined, and it is of the greatest importance to measure the motions of the mind. I beg your majesty to measure it." (Mencius, circa 335 BC, quoted by Lau (1970)). It is apparently a small step between weighing a potato, and understanding the human psyche. Worse still, however, the early attempts at using numbers in landscape work demonstrated significant failings, even on their own terms. Hamill (1995) considered that these faults comprise

- the incorrect use of numbers derived from a place in a classification,
- the incorrect use of numbers to stand for words,

- the use of spurious numbers in simple mathematical operations,
- the use of bad data in complex mathematical and statistical operations,
- the use of data that do not satisfy the requirements of the model,
- the use of numbers to derive, or demonstrate, meaningless, spurious or useless concepts, and
- the use of concepts without adequate operational definitions.

Aside from these problems, there was, and still is, an inbuilt reluctance to ceding responsibility to an inhuman system. Landscape appreciation, we are often told, cannot be separated from the human experience, affected as it is by indeterminable factors such as culture, history and myth.

The result is that IT applied to landscape is still seen by decision and policy makers as an interesting academic exercise, but one step removed from the real world. Surprisingly, this is true even for the area where IT can claim to have the greatest influence, in Visual Impact Analysis (VIA), and in monitoring. The chapter briefly describes the role of IT in landscape planning, with different purposes and values, and also analyses the different degrees of success.

15.2 IT and Visual Impact Analysis

In 1998 the Countryside Council for Wales (CCW) commissioned the Macaulay Land Use Research Institute (MLURI), to undertake research on the best methods of assessing the cumulative visual impact of wind turbines. Some of the impetus for this research came from the statements in an inspector's public inquiry report. The inspector said that he was grateful for the computer analysis and visualizations, but felt that, on a site like this, the problem was too complicated for computers, and best resolved by a site visit. It is hard to understand how this can be true in, for example, assessing the Zone of Visual Influence, which is the area over which a development is visible. Part of the agenda for the research was to establish the credibility of IT methodologies, so that a measure of impact could be established that could be used across the whole of Wales. The need to standardise evaluations, whether of landscape impact or landscape quality, across a region or country, is one of the major arguments for using IT. It is fair to say, however, that considerable progress has been made in developing IT based VIA, and having it accepted by decision-makers (Miller *et al.*, 1997).

For monitoring also, IT seems to offer advantages. Since the landscape is inherently dynamic, landscape planning means managing change. The potential for detecting and analysing land cover change through remote sensing is obvious and can be taken to high levels of sophistication. Precision agriculture links the monitoring of crops, for example using satellite data, to the analysis of the reflection patterns of near-infrared radiation, and shows with the aid of Global Positioning Systems which fields need water, fertiliser or pesticide. Tree species can be distinguished, and their health monitored, by reading their signature wavelength. Such techniques are also invaluable for the landscape planner's purpose, namely identifying the impact of development on the environment, and detecting, for example, the effect of lowering the water table when a quarry is extended. Although the take-up of such techniques is surprisingly slow, their potential is widely recognised.

15.3 Landscape Categorisation and Evaluation in Wales: LANDMAP

Resistance to IT in wider scale landscape endeavours is, however, endemic, in particular in the fields of categorisation and evaluation. In CCW a landscape evaluation and management system called LANDMAP is being developed and is still at the consultation/draft stage. This was originally conceived as a paper based exercise, because, it was said, there were advantages in not using geographical information systems (GIS); it was too complicated and too technical! Three pilot exercises were launched to test the methodology, and it was discovered that without GIS the combination and overlay of the various elements that make up the landscape in the system were simply not practical, or even physically possible. The situation now is that all Local Authorities are encouraged to use LANDMAP to provide the national coverage needed. In theory the Local authorities can do this with paper maps instead of GIS, in practice they are seriously discouraged from trying.

It may be thought that this represents progress in the use of IT in landscape assessment. Unfortunately, however, the methodology was designed to cope with paper and acetates. It not only fails to take advantage of GIS capabilities, but in many ways severely diminishes the analytic properties that IT and GIS can have. Two examples will suffice. One is that the Phase 1 survey of habitats for the whole of Wales, which uses 16 different classes of land cover, is not used as raw data. An 'aspect specialist' was required to divide the vegetation into 'important' and 'not important' classes. This approximated to 'semi natural' and 'non-natural' areas. Similarly, although there is a detailed GIS database being developed further by the Forestry Commission, LANDMAP is unable to recognise whether an area is an ancient woodland or a Sitka spruce plantation, or when the trees are due to be felled, a crucial opportunity for managing change. The forestry aspect of LANDMAP is split into the 'important' and 'not important' and the criterion for the split was whether the area was over or under 50 ha. This will have to be modified, but the methodology will not allow the data to be used without some form of simplification, and the opportunities offered by holding data in digital format will be lost.

In LANDMAP the raw data has been interpreted and stored as information, and can no longer be revisited and interrogated with new questions. The temporal comparisons possible with GIS cannot be carried out. In a sense this is not GIS at all, but simply a database of information and opinion attached as attribute data to polygons on a map, which are themselves not derived by computerised spatial analysis, but arbitrarily hand drawn and then transferred to computer. GIS is not used for analysing data but for storing information; GIS is not used for spatial analysis but as a cartographic tool. Of course all of these arbitrary polygons and expert opinions are immediately available at the touch of the computer's mouse, but nevertheless it is a system that has all the surface flash of GIS, with little of the underlying functionality.

15.4 Landscape Planning Tools in England and Scotland

The CCW system is principally one that stores information about the landscape. That is, it is used for landscape inventory. The initiatives taken in England and Scotland have different levels of use of IT. The system in Scotland, developed by Scottish Natural Heritage (SNH), uses little IT in its derivation, although outputs are being digitised for subsequent use in GIS systems. In this case IT does not confuse the methodology. SNH rely on landscape professionals, working within a tight brief and using an expert consensus, to make their assessment of categories repeatable. It is by no means self evident, however, that these judgements will replicate or reflect the judgements of the public, particularly where categorisation involves judgement of quality. Assessment of landscape character and quality, unlike ecology, is a democratic judgement. As Hawkins (1964) said, " We link together our various perceptual spaces whose contents vary from person to person and from time to time, as parts of our public spatio temporal order ...".

Landscape derives from the interaction of landscape resources and human perception, on the values, past experiences and socio-cultural conditioning of the observer. To a planner it is the public's perception of landscape quality that is important, and it follows that it is the public's perception of landscape character that is important too.

It is important to note that IT offers a new opportunity for the general public to take a greater part in assessing directly the effect of new development on the landscape. The VIA includes, for example, web based virtual reality modelling, which allows the easy distribution of visual impact modelling to the general public. Involving the public is important because there is in any judgement of landscape an underlying aesthetic, value system, not necessarily shared by non-professionals. For CCW the aesthetic is ecological, and works, such as those of Capability Brown or the building of a new golf course, are seen from this perspective as totally negative impacts on the landscape. Bogs, swamps and mires are however, highly valued. It is apparent that our cognitive assessment rapidly affects our affective assessment, in other words what we know about a landscape affects the way we feel about it. SNH seeks to avoid this thorny problem by sticking to categorisation rather than evaluation (Hughes and Buchan, this volume). If this is so, and if the effects of individual human judgement are to be minimised, it is difficult to understand why computer classification and categorisation techniques cannot be used, combined with a formal public perception exercise.

A computerised classification technique has been developed, for example, in New Zealand (Brabyn, 1996). Given the right data, it is estimated that it would take only three days of machine time to complete an equivalent process for an area as large as Wales. Compared to conventional landscape analysis, the investment required is relatively small, but there has been considerable reluctance to employ these techniques, even as an adjunct to our conventional processes.

English Nature has produced the Character of England Map, which has been widely criticised as being oversimplified and too coarse for practical use by Local Authorities. From a technical perspective, the derivation of the map is underpinned by some very interesting spatial analysis. This data are not summarised and partially lost as in CCW's LANDMAP, but they are hidden. The underlying data were not widely circulated since they might distract from the main output, the map.

15.5 The Requirements of Landscape Planning

Why should information technology be used if the national conservation/landscape bodies find it unnecessary or irrelevant? The arguments for its use in VIA or monitoring have been made in section 15.2, but for classification, evaluation and planning there are equally valid reasons for its wider adoption.

A major issue, particularly for national organisations, is that landscape classification or evaluation should be repeatable independently of the original assessment, either over time or in different areas. It should also be transparent in a way in which individual assessment is not. The problem of impartiality and transparency can be solved to an extent by forcing the experts to divide the landscape into restricted categories on the basis of strict criteria. To make these repeatable, however, they may have to be grossly simplified; the computer may paradoxically be able to use more categories and more complex criteria.

It is unlikely that aggregating the various judgements of individual experts will allow for either the calculation of uniqueness or the determination of variety. It may also prove difficult to monitor and thus manage change. Furthermore, for a national organisation with responsibility for individual sites the ability of GIS to allow aggregation and disaggregation of data at national, regional and site levels, is of vital importance.

The advantages that GIS can bring are

- to enable the definition of landscape types and forms as well as landscape areas at a national level, and in a way repeatable over time;
- to provide a database system for storing ecosystem data on a national level;
- to detect and monitor landscape change on a macro scale;
- to allow aggregation and disaggregation of data (at regional, landscape, site scales);
- to locate landscape study areas and ecologically sensitive areas;
- to support spatial analysis of ecological distributions (for example the Natura 2000 network of sites);
- to provide input data/parameters for ecosystem modelling and environmental assessments; and
- to enable input and output from remotely sensed data.

The contribution of traditional landscape techniques should not be undervalued, particularly at local and site scales. But for regional and national assessments, the advantages of computer based techniques would seem invaluable.

15.6 Conclusions

There is an apparent unwillingness on the part of end users, public authorities and landscape professionals to embrace the new techniques being developed by universities and commercial institutions. Suggesting possible solutions is inevitably more problematic. There is a strong case for planning and public authorities to take a fresh look at the current range of IT techniques. It is foolish to suppose that judgements of value could or should be made by computer analysis. Nevertheless, IT has particular advantages that cannot be matched by traditional assessment methods, for example in using the vast quantity of remote sensing data now available, or in making landscape assessment a more democratic process.

References

Brabyn, L. 1996. Landscape classification using GIS and national databases. *Landscape Research*, **21**, 277-300.

Hamill, L. 1985. On the persistence of error in scholarly communication; the case of landscape aesthetics. *Canadian Geographer*, **29**, 270-273.

Hawkins, D. 1964. *The Language of Nature.* San Francisco, Freeman.

Lau, D.C. 1970. *Mencius.* New York, Penguin Books.

Miller, D.R., Wherett, J. and Tan, H. 1997. *Assessing the Cumulative Impact of Wind Turbines.* Bangor, Countryside Council for Wales.

16   MODELLING THE VIEW: PERCEPTION AND VISUALIZATION

Ian D. Bishop

Summary

1.   Landscapes need to be spatially, quantitatively differentiated to be properly included in decision making.
2.   Quantitative landscape research has been working towards this end for over 20 years.
3.   The range of research areas has expanded with growing computational capabilities.
4.   Some of the new development areas are reviewed.
5.   The new developments need to be used in the context of basic landscape issues.

16.1 Introduction

There is a point of view (e.g. Carlson, 1984; Hargrove, 1989), sometimes referred to as 'Positive Aesthetics', that

- everything in nature has overall positive aesthetic value, and
- this positive value cannot be rated, ranked or otherwise distinguished in its equality.

As Godlovitch (1998, p. 195) points out " If Positive Aesthetics resists the ranking of nature's things, then Positive Aesthetics alone cannot discriminate between the aesthetic value of aspects of nature. But if Positive Aesthetics cannot thus discriminate, it cannot offer anything decisive in conservation deliberations where choice of a favoured site is forced upon us, as it always is." This is also the underlying contention of this paper. If aesthetic values are to be considered in environmental or landscape planning then they need to be evaluated: and the earlier in the process the better. Effective landscape planning requires all of the relevant costs and benefits, inputs and outcomes, to be considered simultaneously in the decision making process. It is not good enough that economic and engineering issues are judged in one round, with environmental concerns dealt with as an afterthought. Integration in the decision making process requires that landscape analysis be quantitative, spatially discriminating, commensurable with other parameters and effectively communicated.

This need for explicit assessment was the basis for much of the research presented at the first major conference on analytical and quantitative approaches to landscape management (Elsner and Smardon, 1979). The research initiatives reported at the 'Our National

Landscape' conference (Elsner and Smardon, 1979), and in contemporary research papers, tended to stand alone. An overall structure of development towards a visual resource management toolbox was perhaps implicit – but not clearly articulated. Over the last 20 years there have been considerable advances in a number of research areas, but the way in which these fit together is not frequently debated. This chapter includes a conceptual pattern of linkages based on landscape aesthetic research from the last few years. The papers from a similar period leading up to 'Our National Landscape' are also considered in the light of this framework to assess what changes of emphasis have occurred since 1979.

One key factor in this change is the rapid development of computer technology. In 1979 technology had only recently moved from job submission by punched cards to interactive terminals linked to the 'main-frame' computer. The very first micro-computers were just appearing. Some had as much (for the day) as 16K of memory. Computer Graphics could be generated by overpunching on a line-printer. On-screen pictures (in green and black) could be attempted on the new micros by use of special 'graphic characters'. Well-heeled laboratories had Tektronix terminals on which rudimentary maps and simple perspectives could be drawn. Computer developments over the last 20 years have created a wealth of new opportunities for the landscape researcher in the areas of both modelling and visualization.

16.2 Research – Twenty Years of Growth

Figure 16.1 is a framework based on research papers from the period 1975 to 1979. Table 16.1 gives some typical titles. During this period there was a clear emphasis on perception - on developing a body of theory about the triggers of landscape preference. This was essentially photograph based. As the theory developed other authors looked to incorporate the lessons learnt into landscape planning processes. The potential for computers to support perception research, the planning process and visualization, was clearly emerging.

A research tree based on publications from the last 5 years can be constructed with the same basic structure of roots and stems, but newly developed research areas have added new branches and foliage (Figure 16.2). Notable among these are the emphasis on visual presentation using advanced computer graphics, the employment of a range of spatial information (geographic information systems and remote sensing), other emerging technologies (artificial intelligence and the Internet) and attempts to integrate these into more comprehensive systems. A list of illustrative titles of papers since 1994 shows the range of new research areas (Table 16.2).

16.3 The Computer as Support Tool

This section examines some specific areas in which new technologies are contributing to our knowledge of, and ability to work with, landscape character and change. In many cases these tools require adaptation in order to work effectively in specific landscape character types. Effective character definition therefore underpins many of the new developments in landscape research. The choice of specific subjects reflects the research opportunities and interests of the author.

16.3.1 Geographic Information System based modelling
The physical elements of the environment determine how it looks. Geographic information

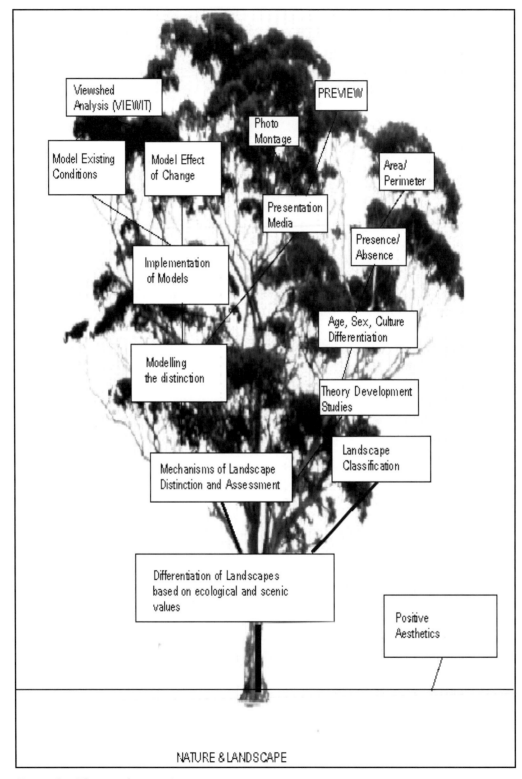

Figure 16.1. The research tree – circa 1979.

Table 16.1. The table lists twelve typical titles of papers published between 1975 and 1979. Key words are highlighted in bold print.

Title	Author(s)
Visual analysis: a **computer-aided** approach to determine **visibility**.	Aylward and Turnbull, 1977
Application of a landscape-preference model to **land management**.	Brush and Shafer, 1975
Psychological factors in landscape appraisal.	Craik, 1972
Measuring landscape aesthetics: the **scenic beauty estimation** method.	Daniel and Boster, 1976
Landscape with **photographs**: testing the preference approach to landscape evaluation.	Dunn, 1976
Appraising the **reliability** of **visual impact** assessment methods .	Feimer *et al.*, 1979
Visual land use compatibility as a significant contributor to **visual resource quality**.	Hendrix and Fabos, 1975
Rated preference and **complexity** for natural and urban visual material.	Kaplan *et al.*, 1972
PREVIEW: **computer assistance** for visual management of forested landscapes.	Myklestad and Wager, 1977
How to **measure** preferences for photographs of natural landscapes.	Shafer and Brush, 1977
Validation of a natural landscape preference model as a predictor of perceived landscape beauty in photographs.	Thayer *et al.*, 1976
The relationship of **observer and landscape** in landscape evaluation.	Unwin, 1975

Table 16.2. In comparison with Table 16.1, this table lists the titles of twelve papers published since 1994. Key words are highlighted in bold print.

Title	Author(s)
The **validity** of computer-generated graphic images of forest landscape.	Bergen *et al.*, 1995
Towards a **virtual reality** interface for landscape visualization.	Berger *et al.*, 1996
Predicting scenic beauty using mapping data and **geographic information systems**.	Bishop and Hulse, 1994
Testing perceived landscape colour difference using the **Internet**.	Bishop, 1997
An **AI** methodology for landscape visual assessments.	Buhyoff *et al.*, 1994
Using **remotely sensed** data in landscape visual quality assessment.	Crawford, 1994
Perspective terrain visualization - a **fusion** of remote sensing, GIS and computer graphics.	Graf *et al.*, 1994
Integration of computerized visual simulation and visual assessment in **environmental planning**.	Lange, 1994
A perceptual evaluation of **computer-based** landscape simulations.	Oh, 1994
SmartForest: A 3-D **interactive** forest visualization and analysis system.	Orland, 1994
The effects of **visual variety** on perceived human preference	Orland *et al.*, 1994
Perceived **scale accuracy** of computer visual simulations.	Watzek and Ellsworth, 1994

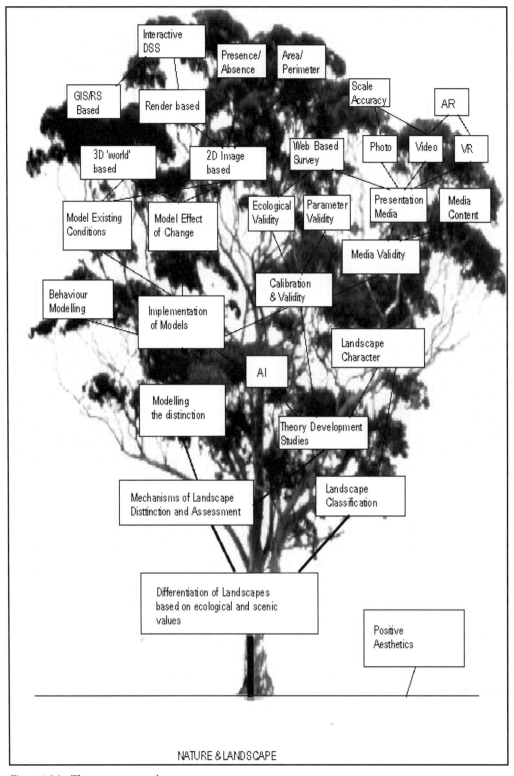

Figure 16.2. The current research tree.

systems (GIS) store data about the physical environment. The logical link seems clear but has not been widely exploited. Is this because landscape is widely treated as a two dimensional entity: something seen as if through a camera lens? Johnson (1979) (quoted in Rolston, 1995) put it this way: "Landscape *per se* does not exist; it is amorphous – an indeterminate area of the earth's surface and a chaos of details incomprehensible to the perceptual system. A landscape requires selective viewing and a frame".

Viewshed analysis (Aylward and Turnbull, 1977) can provide that frame by identifying what can be seen – within a specific arc – from any particular viewpoint. It can also give an indication of what might be seen through extension of the basic algorithm (Fisher, 1996). Computed visibility has been exploited in several landscape quality studies (e.g. Hadrian *et al.*, 1988; Steinitz, 1990; Bishop and Hulse, 1994; Xiang, 1996). The complexity of models has ranged from the simple ability to see a particular landscape feature or a set of features to exploration of the mysterious 'black-box' of artificial neural nets (Bishop, 1996).

Other studies have also used GIS as a basis for visual modelling without drawing on viewshed analysis. This can be a dangerous approach, because of the chaos of detail and absence of frame, but can be justified when undertaken at a regional level with large analysis zones (usually grid cells). Mendel and Kirkpatrick (1999), for example, mapped the aesthetic values of Tasmania on a 10 km square grid. Much earlier the US land management agencies were using presence/absence data to map scenic values (Bureau of Land Management, 1975).

The reasons why the viewshed approach has not received more widespread attention are unclear. The predictive capabilities of regression models based on elements in the viewshed are similar to those based on quantitative photographic analysis. The problem may be that the GIS data are usually insufficiently detailed to be trusted as a full description of the landscape. Vegetation, for example, is generally not sufficiently well mapped in either height or density. Also, the data will not always extend to the boundaries of the viewshed, thus omitting potentially significant landscape features (such as distant hills). Finally, the processing is typically slow. Running viewshed analysis on every point of a fine raster is still a very time consuming process. A potential solution to this last concern is to use the Z-buffer hidden surface algorithm that is integral to three dimensional rendering. This determines not only what can be seen in the view but also how far away it is. With 3D graphics accelerator cards becoming a common feature of personal computers the calculation of viewsheds can become hardware rather than software based.

16.3.2 *Virtual and Augmented Reality*

In the computer graphics community Virtual Reality (VR) is a high-tech operation involving head-mounted displays (HMDs), Data Gloves, and for the really fortunate a CAVE (Cave Automatic Virtual Environment), within which the user is surrounded by projected images. Among landscape researchers the term VR is often applied to the output of more modest apparatus. A virtual reality has been defined as an environment created by the computer in which the user feels present. Offerings may range from landscape simulation generators such as VistaPro (Berger *et al.*, 1996), to animation (Herrington *et al.*, 1993) to highly interactive environments (Graf *et al.*, 1994; Lange, 1994) and to the full immersive condition of a CAVE.

The reason for applying VR to landscape may be simply to communicate the qualities of a specific plan or design, to provide for interactive manipulation of design elements (Orland, 1997) or to undertake experiments in perception (Bishop and Rohrmann, 1995). In general, the greater interactivity demands lower detail and hence a lower level of visual realism. Consequently a trade-off must be made. This raises the question - how real is real enough? This, in turn, makes the undertaking of validity studies a significant issue. Validity studies in relation to landscape presentation media have been going on for a long time – beginning with photographs (e.g. Shuttleworth, 1980), moving to video (Vining and Orland, 1989) and more recently dealing with computer graphics (Oh, 1994; Watzek and Ellsworth, 1994; Bergen *et al.*, 1995). To date, the immersive option, and whether people respond in a 'natural' way when wearing an HMD, has not been widely explored.

One of the reasons why realism in landscape simulation is hard to achieve is that the landscape is highly complex. In computer graphics terms, it is made up of many surfaces and textures. This makes the use of augmented reality (AR) a promising option. AR can be defined as a process in which the real world is enhanced or augmented with additional information generated from a computer model. In this situation we can use imagery of the existing conditions directly and simply superimpose computer generated imagery of those elements of the landscape which are going to change. This paradigm has been exploited for a long time in the form of photomontage and image manipulation, but the new technology offers the prospect of both animation and real-time exploration using the AR principle. There are three main issues to be dealt with in producing successful AR: registration, occlusion and lighting.

Registration from a single viewpoint has been successfully dealt with by many involved in landscape simulation (e.g. Shang, 1992). The same principles can be applied as the viewpoint moves but exact details of viewer's position and orientation must be known or be able to be deduced. Occlusion is the process of hiding those parts of the computer model which should not be visible because of an intervening landscape element. This requires that the existing conditions also be fully modelled – along with the changes. When the whole model is rendered, the existing elements are painted a uniform flat colour which acts as the chroma key for blending the image with existing video input (Plate 22). Lighting is a complex issue in environments with multiple light sources (see Fournier *et al.*, 1993), but is somewhat easier in the landscape situation because there is typically a single light source – the sun. The computer model must include a light source which illuminates the introduced elements in the same way (direction, brightness, etc) as the existing conditions.

16.3.3 *Web based survey*

A common difficulty for perception researchers – whether by photograph or computer graphic techniques – is to find a suitable and sufficient audience to evaluate the stimuli. A solution, which has been explored recently (Bishop, 1997; Wherrett, 1997), is to conduct the survey using the World Wide Web (WWW). Wherrett (1999) reviewed the issues involved in conducting such a survey. These may include concerns about the representativeness of the sample, the comparability of presentation of the stimuli and image quality since it is necessary to keep images to a small file size for reasonable access times.

Questions explored in this way include the validity of a visual colour contrast formula (Bishop, 1997) and scenic quality assessment (Wherrett, 1997). Other examples found on the web include assessment of affective response to safety issues in the landscape (Plate 23) and tourism assessment for a wildlife park. Use of this facility can be expected to grow rapidly as both computer power and web bandwidth increase. The possibility also exists to give people interactive access to three dimensional environmental models using protocols such as the Virtual Reality Modelling Language (VRML) or similar web-based 3D modelling standards.

16.3.4 Behavioural Modelling

In addition to knowing what people like in a landscape and why, there are also advantages for the landscape planner or manager in gaining insight into how people behave under particular environmental conditions. The manger may be interested in behaviour on an aggregated basis – how many people will take this path, how long will they stay at this location, will they be satisfied with the experience? Collective behaviour is, however, made up from the behaviour of many individuals. This is the basis of the emerging area of Autonomous Agent modelling. The concept is that each agent (individual visitor) has their own set of rules that describe their behavioural (and potentially emotional) response to particular sets of circumstances. After placing a number of agents in a site the simulation is run with each agent acting autonomously. After a period of model operation a picture emerges of collective behaviour. Individual or agent-based models have developed in the 1990s as a popular approach to modelling spatially-explicit environmental phenomena. Much of the initial development was based on creatures with much simpler behavioural rules than human tourists. These rules involved responses to the immediate environment and other nearby agents. The agents were not given goals or any ability to learn about their environment. More recently researchers have introduced complex agents whose behaviour develops (or emerges) as each agent learns more about its surroundings and which are capable of adapting complex goals. Several software products to support agent modelling have been developed (e.g. Hiebeler, 1994; Forrest and Jones, 1994).

Gimblett (1998) and Bishop and Gimblett (1998) have described how agent modelling can be linked to a GIS that provides the base environmental information, and how rudimentary behavioural rules can be derived from visitor surveys. An agent-based approach was then used to model movement along existing park trails and the levels of (unsatisfactory) interaction between different recreator types (e.g. hikers and mountain bike riders). In the future we may use virtual reality environments to test behavioural responses to alternative landscape options and hence derive behavioural rules under controlled experimental conditions.

16.4 Integrated and Interactive Systems

These computational contributions to understanding and modelling responses to landscape become increasingly valuable if they can become part of a wider process of decision support. The premise of the introduction was that we need an ability to quantify landscape and landscape changes in order that decisions are not made without due consideration to the landscape values affected by them. It was further argued that there is extra advantage if the landscape modelling can be integrated with the software used in commercial or governmental sectors to analyse engineering, economic or even social factors. Developments in computing are making the integration of disparate software modules considerably easier (Yates and Bishop, 1998a, 1998b). Examples linking GIS to visualization in the landscape context include Liggett and Jepson (1995) and Tang *et al.* (1998). Increases in computer performance, and the demands of users, make it possible and appropriate for integrated systems also to perform expeditiously enough to be interactive.

In an example linking different types of model, Bishop and Gimblett (1998) suggested that the outputs of a tourist behavioural model become inputs to a comprehensive model of tourism effect on landscape using GIS based modelling. The effect of particular management decisions will influence visitor movements. These in turn will determine the environmental effects of tourism. Figure 16.3 shows how management options are processed through the system to generate visitor satisfaction levels and environmental impact levels.

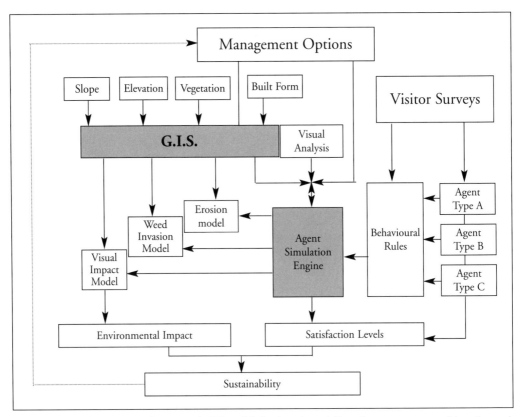

Figure 16.3. The concept of integrated GIS and behavioural modelling for management of heavily used landscapes. The right-hand side of the figure is the agent-based behavioural modelling research of Gimblett (1998) as described in Section 16.3. The left-hand side is GIS based impact modelling.

16.5 Conclusion

Landscape research is in general not of sufficient prominence to drive technology development. The differences between the landscape research trees from the 70s and from the 90s (Figures 16.1 and 16.2) make it clear that researchers have not been slow to adopt new capabilities as they have emerged. So, is landscape research dealing with the questions that really need to be answered, or are researchers simply playing with new toys because they are fun and generate publications?

There is likely to be truth on both sides. Certainly we are finding out a great deal about ways to adapt emerging technologies to landscape issues but, at the same time, there are many fundamental questions of human-environment interaction to which we do not have answers.

Those engaged in computer based landscape research will presumably answer that they are seeking to use the emerging technological opportunities to advance fundamental knowledge as well as creating new opportunities to use the knowledge we already have. Others may be inclined to the view that the complexity of landscape and landscape response demands deeper thinking than is implied in a preference survey, a regression equation or a computer simulation.

My expectation is that these views will converge and that the full potential of computer visualization to develop and support our knowledge of landscape perception will be realised. In this way we may model the view – not only as a picture in a frame, but as a component of human attitudes to, behaviour in, and respect for the environment.

References

Aylward, G. and Turnbull, M. (1977). Visual analysis: a computer-aided approach to determine visibility. *Computer-Aided Design, 9*, 103-108.

Bergen, S.D., Ulbricht, C.A., Fridley, J.L. and Ganter, M.A. (1995). The validity of computer-generated graphic images of forest landscape. *Journal of Environmental Psychology, 15*, 135-146.

Berger, P., Meysembourg, P., Sales, J. and Johnston, C. (1996). *Towards a virtual reality interface for landscape visualization.* Third International Conference/Workshop on Integrating GIS and Environmental Modelling., Santa Fe, New Mexico. Published only as a CD-ROM.

Bishop, I.D. (1996). Comparing regression and neural net based approaches to modelling of scenic beauty. *Landscape and Urban Planning, 34*, 125-134.

Bishop, I.D. (1997). Testing perceived landscape colour difference using the Internet. *Landscape and Urban Planning, 37*, 187-196.

Bishop, I.D. and Gimblett, H.R. (1998). *Modelling tourist behaviour: geographic information systems and autonomous agents.* Proceedings Tourism and Culture for Sustainable Development, Athens. Published only as a CD-ROM.

Bishop, I.D. and Hulse, D.W. (1994). Predicting scenic beauty using mapping data and geographic information systems. *Landscape and Urban Planning, 30*, 59-70.

Bishop, I.D. and Rohrmann, B. (1995). Visualization tools: their role in understanding human responses to the urban environment. In Wyatt, R. and Hossain, H. (Eds) *Urban Planning and Urban Management.* Dept of Geography and Environmental Studies, University of Melbourne. 1: 313-322.

Brush, R.O. and Shafer, E.L. (1975). Application of a Landscape-Preference Model to Land Management. In Zube, E.H., Brush, R.O. and Fabos, J.G. (Eds). *Landscape Assessment: Values, Perceptions and Resources,* Pennsylvania, Halstead Press, 168-181.

Buhyoff, G.J., Miller, P.A., Roach, J.W., Zhou, D. and Fuller, L.G. (1994). An AI methodology for landscape visual assessments. *AI Applications, 8*, 1-13.

Bureau of Land Management (1975). Visual Resource Management. *BLM Manual* 6300. US Department of the Interior, Washington DC.

Carlson, A. (1984). Nature and positive aesthetics. *Environmental Ethics, 6*, 5-34.

Craik, K.H. (1972). Psychological factors in landscape appraisal. *Environment and Behavior, 4*, 255-266.

Crawford, D. (1994). Using remotely sensed data in landscape visual quality assessment. *Landscape and Urban Planning, 30*, 71-92.

Daniel, T.C. and Boster, R.S. (1976). Measuring landscape aesthetics: the scenic beauty estimation method. *USDA Forest Service Research Paper RM-167.* Rocky Mountain Forest and Range Experiment Station, Forest Service, US Department of Agriculture.

Dunn, M.C. (1976). Landscape with photographs: testing the preference approach to landscape evaluation. *Journal of Environmental Management,* **30,** 47-62.

Elsner, G.H. and Smardon, R.C. (Eds) (1979). Our National Landscape: a conference on applied techniques for analysis and management of the visual resource. *USDA Forest Service* PSW-35. Berkeley, CA.

Feimer, N.R., Craik, K.H., Smardon, R.C. and Sheppard, S.R.J. (1979). Appraising the reliability of visual impact assessment methods. In Elsner, G.H. and Smardon, R.C. (Eds). *Our National Landscape: a conference on applied techniques for analysis and management of the visual resource,* USDA Forest Service PSW-35. Berkeley, CA.

Fisher, P. (1996). Extending the Applicability of Viewsheds in Landscape Planning. *Photogrammetric Engineering and Remote Sensing,* **62,** 1297-1302.

Forrest, S. and Jones, T. (1994). Modeling complex adaptive systems with Echo. In Stonier, R.J. and Xing Huo Yu (Eds) *Complex systems, mechanism of adaptation,* IOS Press, 3-20.

Fournier, A., Gunawan, A.S. and Romanzin, C. (1993). Common illumination between real and computer generated scenes. Proceedings of Graphics Interface, Toronto, 254-262.

Gimblett, H.R. (1998). Simulating Recreation Behaviour in Complex Wilderness Landscapes Using Spatially-Explicit Autonomous Agents. Unpublished PhD Thesis, University of Melbourne, Parkville, Australia.

Godlovitch, S. (1998), Valuing nature and the autonomy of natural aesthetics. *British Journal of Aesthetics,* **38,** 180-197.

Graf, K.C, Suter, M., Hagger, J., Meier, E., Meuret, P. and Nuesch, D. (1994). Perspective terrain visualization - a fusion of remote sensing, GIS and computer graphics. *Computers and Graphics,* **18,** 795-802.

Hadrian, D.R., Bishop, I.D. and Mitcheltree, R. (1988). Automated mapping of visual impacts in utility corridors. *Landscape and Urban Planning,* **16,** 261-283.

Hargrove, E. (1989). *Foundations of Environmental Ethics.* Englewood Cliffs NJ, Prentice Hall.

Hendrix, W.G. and Fabos, J.G. (1975). Visual land use compatibility as a significant contributor to visual resource quality. *International Journal of Environmental Studies,* **8,** 21-28.

Herrington, J., Daniel, T.C. and Brown, T.C. (1993). Is motion more important than it sounds? The medium of presentation in environmental research. *Journal of Environmental Psychology,* **13,** 283-291.

Hiebeler, D. (1994). The Swarm Simulation system and individual-based modelling. *Proceedings of Decision Support 2001. Resource Technology '94 Symposium, American Society for Photogrammetry and Remote Sensing,* **1,** 474-494.

Johnson, H.B. (1979). The framed landscape. *Landscape,* **23,** 26-32.

Kaplan, S. and Kaplan R.W. and Wendt, J.S. (1972). Rated preference and complexity for natural and urban visual material. *Perception and Psychophysics,* **12,** 354-365.

Lange, E. (1994). Integration of computerized visual simulation and visual assessment in environmental planning. *Landscape and Urban Planning,* **30,** 99-112.

Liggett, R. and Jepson, W. (1995). An integrated environment for urban simulation. *Environment and Planning B (Planning and Design),* **22,** 291-305.

Mendel, L.C. and Kirkpatrick, J.B. (1999). Assessing temporal changes in the preservation of the natural aesthetic resource using pictorial content analysis and a grid-based scoring system – the example of Tasmania. *Landscape and Urban Planning,* **43,** 181-190.

Myklestad, E. and Wager, J.A. (1977). PREVIEW: computer assistance for visual management of forested landscapes. *Landscape Planning,* **4,** 313-331.

Oh, K. (1994). A perceptual evaluation of computer-based landscape simulations. *Landscape and Urban Planning*, **28**, 201-216.

Orland, B. (1994). SmartForest - A 3-D interactive forest visualization environment. *Proceedings Decision Support 2001- Resource Technology '94, American Society for Photogrammetry and Remote Sensing, Washington, DC*, 181-190.

Orland, B.A. (1997) Forest visual modeling for planners and managers. *Proceedings ASPRS/ACSM/RT'97 Seattle, American Society for Photogrammetry and Remote Sensing, Washington, DC*, **4**, 193-203.

Orland, B., Radja, P., Larsen, L. and Weidemann, E. (1994). The effects of visual variety on perceived human preference. *Proceedings, Society and Resource Management*. Fort Collins, Colorado.

Rolston, H. (1995). Does Aesthetic appreciation of landscapes need to be science-based? *British Journal of Aesthetics*, **35**, 374-386.

Russell, J.A. and Lanius, U.F. (1984). Adaptive level and the affective appraisal of environments. *Journal of Environmental Psychology*, **4**, 119-135.

Shafer, E.L. and Brush, R.O. (1977). How to measure preferences for photographs of natural landscapes. *Landscape Planning*, **4**, 237-256,

Shang, H.-D. (1992). A method for creating precise low-cost landscape architecture simulations - combining computer aided design with computer video-imaging techniques. *Landscape and Urban Planning*, **22**, 11-16.

Shuttleworth, S. (1980). The use of photographs as an environmental medium in landscape studies. *Journal of Environmental Management*, **11**, 61-70.

Steinitz, C. (1990). Toward a sustainable landscape with high visual preference and high ecological integrity: the loop road in Acadia National Park, U.S.A. *Landscape and Urban Planning*, **19**, 213-250.

Tang, H., Bishop, I.D. and Yates, P.M. (1998). Towards an integrated interactive visualization system for forest management. Proceedings of *Resource Technology* 97. Beijing, China Forestry Publishing House, pp. 250-258.

Thayer, R.L., Hodgson, R.W., Gustke, L.D., Atwood, B.G. and Holmes, J. (1976). Validation of a natural landscape preference model as a predictor of perceived landscape beauty in photographs. *Journal of Leisure Research*, **8**, 292-299.

Unwin, K.I. (1975). The relationship of observer and landscape in landscape evaluation. *Transactions of the Institute of British Geographers*, **66**, 130-133.

Vining, J. and Orland, B. (1989). The video advantage: a comparison of two environmental representation techniques. *Journal of Environmental Management*, **29**, 275-283.

Watzek, K.A. and Ellsworth, J.C. (1994). Perceived scale accuracy of computer visual simulations. *Landscape Journal*, **13**, N1(Spring), 21-36.

Wherrett, J.R. (1997). Natural scenic landscape preference: predictive modelling and the world wide web. *Proceedings of the International Conference on Urban, Regional, Environmental Planning and Informatics to Planning in an Era of Transition, Athens*, 775-794.

Wherrett, J.R. (1999). Natural landscape scenic preference: techniques for evaluation and simulation. Unpublished PhD thesis, Robert Gordon University, Aberdeen.

Xiang, W.-N. (1996). A GIS based method for trail alignment planning, *Landscape and Urban Planning*, **35**,11-23.

Yates, P.M. and Bishop, I.D. (1998a). Automatic Generation of Software Supporting the Integration of GIS and Modelling Systems. GeoComputation '98, Bristol. Published only as a CD-ROM.

Yates, P.M. and Bishop, I.D. (1998b). The Integration of Existing GIS and Modelling Systems with Urban Applications. *Computers, Environment and Urban Systems*, **22**, 71-80.

17 HISTORIC LAND USE ASSESSMENT PROJECT

Piers Dixon, Lynn Dyson Bruce, Richard Hingley and Jack Stevenson

Summary

1.   The origins of Historic Land Use Assessment (HLA) lie in the developing practice of Landscape Character Assessment (LCA). LCAs have generated a new and more informed approach to landscape issues. They have also been created with particular purposes and objectives in mind and an assessment of these documents by cultural heritage managers indicates that the mapping scale at which they are undertaken does not enable historical and archaeological information to be adequately identified.

2.   This historical dimension is important as an aid to our understanding of the processes behind the formation of the current landscape. The landscape contains a record of historic events, and it is important to understand this to improve management by informing the wider landscape debate.

3.   Historic Scotland (HS) and the Royal Commission on the Ancient and Historical Monuments of Scotland (RCAHMS) have formed a partnership to develop a HLA methodology for Scotland within a Geographical Information System in order to enable this understanding of the historic landscape. This chapter reports some of the results from a pilot HLA carried out between 1996 and 1998 in six areas of Scotland. It demonstrates regional variation in the types of historic land use patterns occurring across Scotland and also chronological development.

4.   HS and RCAHMS are carrying out further work on HLA during 1998 and 1999 both to help inform the National Parks debate and to develop the methodology further.

17.1 Objectives of Historic Land Use Assessment

The Historic Land Use Assessment (HLA) Pilot Project was established by Historic Scotland (HS) and the Royal Commission on the Ancient and Historical Monuments of Scotland (RCAHMS) to explore the viability of creating a method of assessing historic land use patterns in Scotland. HLA is built using the RCAHMS Geographical Information System (GIS), which currently uses GenaMap software from Genasys Ltd held on a HP Unix server.

The origins of HLA lie in the developing practice of Landscape Character Assessment (LCA), which has generated a new and more informed approach to landscape issues (Scottish Natural Heritage, 1998). Scottish Natural Heritage (SNH) has initiated a national programme of LCA with particular purposes and objectives in mind. However, an assessment of the LCA regional reports by cultural heritage managers indicates that the mapping scale at which they are undertaken does not enable historical and archaeological information to be used to its full potential. This historical dimension is important as an aid to our understanding of the processes behind the formation of the current landscape and is an important dimension of landscape character.

Recently, developments have taken place in the field of Historic Landscape Character Assessment (HLCA) elsewhere in the United Kingdom. The Scottish methodology has been informed by work in Cornwall, which mapped and characterised the patterns of fields (Herring, 1998), but differs from the *Register of Landscapes of Outstanding Historic Interest in Wales* in focusing on a database approach rather than selected landscapes (CADW, 1998).

The landscape contains a record of prehistoric and historic events, and it is important to understand this to inform the wider landscape debate. HS and RCAHMS therefore decided to form a partnership in 1996 in order to develop and test an appropriate mapping methodology for Scotland, using GIS technology.

The main value of HLA lies in its potential to enable the more detailed input of built heritage interests into the management of landscape change, although it has been found to have certain other potential uses. Historic land use is mapped in order to provide a body of information that will allow priorities to be drawn within wider landscape management, which give appropriate weight to the historic dimension of the landscape. HLA can produce multiple maps based on different sets of criteria. Its full value can most easily be appreciated through the use of the GIS system but this chapter provides a summary of its potential and illustrates the potential of the approach using some of the results of the pilot assessments. Discussion focuses upon

- regional and chronological patterning, and
- managing landscape change.

A more comprehensive description, incorporating greater detail of the methodology and more examples of the application of HLA, will be found in Dyson Bruce *et al.* (1999).

17.2 Methodology

17.2.1 The relationship of HLA to LCA

There is a fundamental difference in approach and methodology between HLA and LCA. HLA maps the impact of people upon the land surface. This methodology exclusively identifies the anthropogenic and historical elements of the landscape at the 1:25,000 scale. LCA is specifically designed to assess the characteristics of the landscape according to a set methodology evolved by SNH based on guidance developed by the Countryside Commission for Scotland and the Countryside Commission. It assesses the landscape at 1:50,000 scale (Countryside Commission, 1993).

HLA and LCA therefore map different elements of the landscape at different scales for different purposes but the approaches would appear to be complementary. HS, RCAHMS and SNH are currently exploring the potential of integrating the results of LCA and HLA.

17.2.2 Methods

The mapping process involves the systematic assessment of topographic Ordnance Survey (OS) maps, archaeological and historical data in the National Monuments Record of

Scotland, the Land Cover of Scotland (Macaulay Land Use Research Institute, 1993), and vertical aerial photographs. The assessment was intended to be a broad-brush exercise, but retaining the topographic detail that would allow the historic land use to be characterised. The smallest scale of topographic data which included field boundaries was the OS 1:25,000 Pathfinder maps. This was adopted as the scale of capture. It had the added advantage that the smallest scale of data used by any of the above sources was 1:25,000, as in the case of the land cover data, and it was close to that of the 1988 aerial photographs that formed the basis of that survey. It had the disadvantage that any feature which is less than one hectare in extent is too small to map at the 1:25,000 scale. This means that many individual archaeological sites, including linear sites such as Roman roads, do not show up on the maps that are produced. Groups of structures have been included, with a yardstick, for example, of at least five shielings or three hut-circles per hectare. Small sites can, however, be recovered and used in conjunction with the HLA through the use of the National Monuments Record of Scotland database on the GIS system.

The current landscape is characterised using the OS map as a base, and, by-and-large, the main source of relict land use types is the National Monument Record of Scotland, but across most of Scotland no recent survey (i.e. since 1985) has been done, and the aerial photographs become more critical as a source. However, the quality of the aerial photographs (time of day, time of year, etc.), and the vegetation cover, directly affect the ability to identify and record relict landscapes. For this reason a validation process that involves ground visits to check unconfirmed sites was built into the project. The information from these sources is collated and mapped by the application of a simple but clearly defined series of historical land use types. For ease of use, two main categories of land use type have been defined as

- current land use types - reflecting historic land use types in current use, which may include types that are in origin several hundred years old, and
- relict land use types - reflecting historic land use types that have been abandoned, but which still leave some trace in the landscape.

A glossary of terms has been built-up that includes 42 current types and 37 relict types of which 12 have current equivalents. Each type has a reference number that is used to tag the polygons on the map and if a polygon includes relict types a composite number is created. Each land use type is mapped by eye on a tracing-paper overlay of the 1:25,000 map. The HLA map is designed so that every part has a current land use type, but only where there are visible or mappable relics of past land use is a relict polygon created. Along with the map a database is also being compiled in Microsoft Access of all the single and multiple types that occur. There is a maximum of three relict types in any given area, although it is conceivable that this could be increased for particularly complex landscapes, and there are 246 variations in the database to date.

The resulting composite map is then entered into a GIS system, using GenaMap, to produce topologically correct maps. The digitising is executed on screen using the OS BasicScale digital map as a base. To speed-up the process of digitising, some data, particularly relating to woodland cover, is pasted in using the MLURI land cover data. The polygons are tagged with the historic land use reference number and attached to an up-to-date copy of the database, so that analyses can be carried out textually. Once entered into the GIS, the complete maps are transferred into *Artemis* - a more user-friendly graphical user interface. The maps may then be combined with other datasets for further interrogation and analysis.

Interpreted data of this sort on relict landscape types, current land use and the survival of field patterns are not available from existing archaeological sources, which makes this a unique resource. However, it must be reiterated that it should be used in conjunction with the National Monuments Record of Scotland database of sites.

17.3 Historic Land Use Assessment application

The value of HLA lies in the potential of integrating the information with other datasets held on GIS and this helps to provide a detailed understanding of the modern landscape.

A number of pilot areas in differing topographical zones were selected to develop and test the methodology (Figure 17.1). The assessment of the pilot project areas not only confirmed the expected regional patterns of current land use types but also showed the extent and range of relict land use types which are of particular archaeological interest. The following section illustrates, with three case studies, some of the results derived from the pilot project areas and demonstrates a range of applications for HLA data.

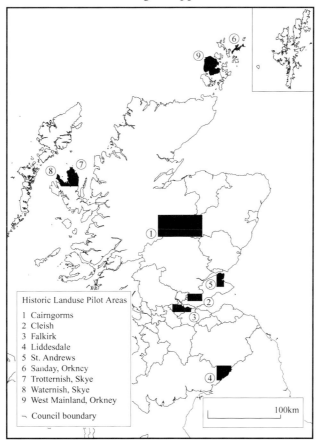

Historic Landuse Pilot Areas

1 Cairngorms
2 Cleish
3 Falkirk
4 Liddesdale
5 St. Andrews
6 Sanday, Orkney
7 Trotternish, Skye
8 Waternish, Skye
9 West Mainland, Orkney

⌐ Council boundary

100km

Figure 17.1. Location map of the nine pilot areas covered by the Historic Land Use Assessment Project.

17.3.1 Regional Patterning

The pilot areas have established that a wide variation in the nature of historic land use types exists across Scotland. This can most easily be demonstrated by considering two detailed case studies. Tables 17.1 and 17.2 show examples of contrasting regional patterns. The first is a lowland agricultural/industrial/urban mix (Table 17.1) and the second an upland moorland/rough-pasture/forestry/agriculture mix (Table 17.2). Although the two landscapes are dominated by modern land use types, both contain important areas of relict types, all of which are threatened by land use changes.

Table 17.1. The distribution of current and relict land use types for Cleish.

Current land use type	Per cent cover	Per cent relict
18th-19th century fields	55	16
Post-World War II prairie	2	3
Restored agricultural land	4	0
Urban	4	4
Rough pasture	11	16
Forestry	10	37
Other current land use types	14	24
Total	100	100

Table 17.2. The distribution of current and relict land use types for Liddesdale.

Current land use type	Per cent cover	Per cent relict
18th-19th century fields	7	16
Rough pasture	42	77
Forestry	20	0
Other current land use types	31	7
Total	100	100

Cleish is an area where rapid landscape change is occurring. The collapse of the deep-mining industry has been accompanied by the expansion of forestry plantations and the development of large-scale opencast mines. On a relatively small scale the Cleish landscape exemplifies the type of developments occurring over much of the Central Belt of Scotland in the wake of the collapse of the coal and iron industries and with the problems facing upland farming.

The Cleish landscape area is dominated by 18th and 19th century rectilinear fields interspersed with intensive urban, industrial and extractive developments (Table 17.1). Relict land use is limited to pre-improvement agriculture, mineral extraction, with prehistoric and later settlement, mainly surviving as crop marks. Policies and parklands (both relict and current) are also a characteristic feature of the landscape, e.g. Blairadam, Cleish and Aldie (Plate 24).

The recent opencast coal mines have had a dramatic impact on the landscape with large areas being subject to the destructive activity of extraction. Indeed, the extent of restored agricultural land is itself a relict artefact of the process. The uplands have been subject to commercial afforestation, which has led to the loss of considerable areas of relict landscapes, including a large part of the policies of Blairadam.

Liddesdale is an area of marginal agriculture. The higher ground is predominantly rough pasture, some of it drained for grazing or managed for grouse shooting, while the more favoured areas of low ground are more intensively farmed and are characterised by 18th and 19th century rectilinear fields. Newcastleton is an example of an Improvement Period planned village, and it is surrounded by a grid of small fields associated with, but not attached to, the houses in the village. Afforestation has made a major impact on the east side of the valley, and the area under trees is still expanding. Extensive prehistoric and pre-improvement relict landscapes survive within the unforested rough pastures (Table 17.2).

In outline, but not in detail, the area is representative of many parts of southern Scotland. It is a landscape that has undergone numerous phases of settlement and agricultural expansion and contraction from the prehistoric period to the present day. During much of the medieval period it lay within a hunting forest, and the impact of this can be seen in the detail of the medieval and post-medieval settlement pattern. Intakes of land that were deforested are defined by externally ditched dykes, which were built successively on to one another as settlement extended into the margins, only to be abandoned in the post medieval period (Plate 25).

17.3.2 *Chronological Depth*

The relict landscapes that survive in Scotland vary in nature and period from area to area, and there is little doubt that, were HLA mapping to be undertaken throughout Scotland, distinctive regional patterns would emerge. The pilot study showed that HLA mapping is particularly well suited to identifying and displaying areas of relict landscape (e.g. Figure 17.2).

The assessment of Liddesdale has demonstrated the survival of extensive and complex relict landscapes, indicating a long sequence of occupation. By selecting from specific fields in the database, a digest of the information can be produced which reveals the extent of three major relict landscape types, i.e. field-systems dating to the prehistoric, medieval and post-medieval periods. These represent the surviving fragments of what were once more extensive systems of archaeologically significant cultivation remains, which are threatened by the spread of commercial forestry. Similar results have been obtained in some of the other pilot areas.

This allows important archaeological landscapes to be mapped, and an assessment made of their extent and spatial relationships. It is also useful for identifying which landscapes have chronological depth embedded within their modern structure. This makes HLA extremely valuable for managing this important cultural resource.

Figure 17.2 shows relict landscapes in Waternish and Trottenish on Skye. These landscapes are made up, almost entirely, of pre-clearance farming systems, including the settlements, infield and shieling huts from which the summer pasture was exploited. There was extensive abandonment of the area from the early nineteenth century (Royal Commission on the

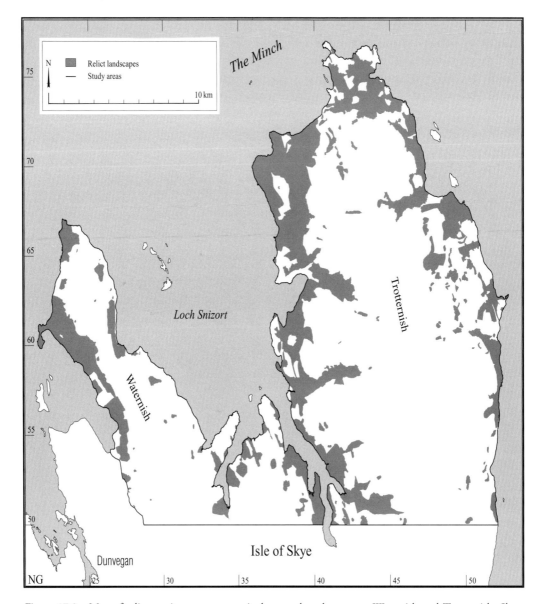

Figure 17.2. Map of relict pre-improvement agriculture and settlement on Waternish and Trotternish, Skye, generated in GIS from data gathered by the Historic Land Use Assessment. The relict pre-improvement farmland has been highlighted to demonstrate the areas of greatest archaeological potential on the two peninsulas.

Ancient and Historical Monuments of Scotland, 1993). These areas have a high archaeological potential and also a relevance to the communities who still live in the area. HLA provides a quick and cost-effective summary of the extent of these relict landscapes.

17.3.3 Managing Landscape Change

The information on prehistoric, medieval, post-medieval and modern landscapes can be fed into the discussions of landscape management that are occurring in Scotland through the use of HLA. HLA mapping has a role to play in the creation of landscape management plans and can be used, in conjunction with Landscape Character Assessments, to add a cultural dimension not readily available from existing archaeological records.

The expansion of commercial forestry and the regeneration of native woodlands are a major concern to archaeologists, and HLA mapping can be used to address the problems associated with the extension of both types of woodland. HLA also has a potential role to play in managing visitor pressure, the allocation of management grants, in strategic planning and the selection of areas for detailed archaeological survey.

17.4 Conclusion

HLA appears, from various demonstrations of the project to a range of audiences, to have value in inputting cultural heritage interests into landscape assessment for a variety of purposes. In addition to providing a regional and chronological understanding of the historic landscape, it can be used to inform landscape management (the location of tree planting, regeneration, houses, industrial development, etc.) and as an aid to the planning of archaeological survey work and management.

HS and RCAHMS are collaborating during 1998 and 1999 to develop a HLA for the proposed National Park in Loch Lomond and the Trossachs, as well as the Cairngorm Partnership area, also a contender for National Park status. The purpose of this exercise is to enable the cultural landscape to be integrated into broader discussions of the management of the landscape of the proposed National Parks. At the same time the exercise will provide a further evaluation of the technique of HLA.

A national coverage would be advantageous as it would provide a national picture of changes in current and relict land use types. It is estimated that it might take twenty person-years to cover the whole of Scotland and HS and the RCAHMS are investigating ways in which HLA might be extended to the whole of the country.

References

CADW (1998). *Register of Landscapes of Outstanding Historic Interest in Wales*. Cardiff, CADW - Heritage in Wales.

Countryside Commission (1993). *Landscape Assessment Guidance*. Cheltenham, Countryside Commission.

Dyson Bruce, L., Dixon, P., Hingley, R. and Stevenson, J.B. (1999). *Historic Landuse Assessment: Development and Potential of a Technique for Assessing Historic Land Use Patterns*. Edinburgh, Historic Scotland and Royal Commission on the Ancient and Historic Monuments of Scotland.

Herring, P. (1998). *Cornwall's Historic Landscape: Presenting a method of Historic Landscape Character Assessment*. Truro, Cornwall Archaeological Unit and English Heritage.

Macaulay Land Use Research Institute (1993). *Land Cover of Scotland 1988*. Aberdeen, Macaulay Land Use Research Institute.

Royal Commission on the Ancient and Historical Monuments of Scotland (1993). *Waternish, Skye and Lochalsh District, Highland Region: an Archaeological Survey*. Edinburgh, Royal Commission on the Ancient and Historical Monuments of Scotland.

Scottish Natural Heritage (1998). *The Landscape Character of Scotland: a National Programme of Landscape Character Assessment*. Perth, Scottish Natural Heritage.

PART FOUR

TOPICS IN LANDSCAPE MANAGEMENT AND CHANGE: REPORTS FROM THE BREAKOUT SESSIONS

18   Planning and Policy Issues

Anne Lumb and David Tyldesley

18.1   Introduction

The discussion session explored the following five themes.

- How can the planning system utilise the Landscape Character Assessment (LCA) programme?
- How can the findings of the LCA be delivered through the planning system?
- How can a more integrated approach to LCA be developed which incorporates the views of the community (stakeholder involvement) and public perception?
- What are the future directions for LCA?
- There was a recognition that application of LCA is still an emerging field, and that there is a need for greater clarity regarding the scope and practical techniques for incorporating LCA into the planning process. There are, however, clear opportunities for the future, as the case study in section 18.2 demonstrates.

18.2 A Case Study: the Stirling and Clackmannanshire Structure Plan Review

Stirling Council, as part of the Structure Plan review process, was charged with allocating a total of between 4,000-6,000 new development units in an area that is of high landscape value. Application of the existing regional LCAs (1:50,000 scale) presented some difficulties because the work was contained in more than one report, the guidelines were varied and there was clearly a need for a more detailed level of information on the landscape resource around settlements. Application of the existing LCAs did, however, enable an initial sift and exploration of all the settlements within the Council area, and the selection of particular areas for further study. The exiting regional LCAs were not sufficiently detailed to consider the areas selected in terms of their landscape capacity, or to inform the housing and development strategy for Stirling in relation to siting, form and mitigation of new development. A more detailed study was therefore commissioned to look at the sifted areas more closely at 1:25,000 and 1:10,000 scales of survey. The work has been used by the Council to influence and inform its locational strategy for new development.

The involvement of the planning authorities in the Steering Group and the development of the regional LCAs have played an important part in encouraging the initial interest in applying the findings of the LCA. This has created a sense of ownership. The existing LCAs, in addition to assisting with the initial selection process for a more detailed study of landscape capacity, will also be useful in reviewing the Indicative Forestry Strategy, and in broad policy development.

18.3 How can the planning system utilise the Landscape Character Assessment Programme?

Landscape Character Assessment is increasingly being recognised as a potentially valuable tool for planners and policymakers. Five years ago, there was no such inventory on the Scottish landscape resource, and its usefulness is perhaps most clearly evidenced by the number of questions now arising in terms of its potential application. All of the Scottish Landscape Character Assessments include guidance for managing pressures for change, and are recognised as having potential to feed into the planning system at many levels. For example,

- National Planning Policy Guidance, where there is scope to take the lead in promoting development, sustaining and enhancing national, regional and local distinctiveness of landscape character;
- Development Plans, where policies and proposals could steer development to locations with capacity to accommodate development and where longer term landscape frameworks and strategies, to increase the capacity of the landscape to accommodate development, could evolve together with policies requiring development to be appropriate to landscape character;
- Relationships with designations, as they form a basis for informing reviews of designations at strategic and local levels; and
- Development control, where there is scope for pro-active use of site development briefs and design guidance, as well as scope for use of conditions and planning agreements.

Examples exist of the practical application of LCA being developed in a number of different areas, for example, the clear and successful applications of LCA in Dumfries and Galloway and in Aberdeen city (Bennett, Nicol and Campbell, this volume). In addition to the ongoing work in Stirling, described in section 18.2, LCAs have provided the basis for

- a new approach to landscape policy in a Consultation Draft Local Plan (in the former) Gordon District,
- new methods of assessing landscape capacity to accommodate built development in the St Andrews Strategic Study (Fife Council),
- assessing landscape capacity to accommodate minerals development (both Dunfermline and Fife Councils),
- assessing landscape capacity in an opencast coal study (Clackmannanshire Council), and
- assessing landscape issues and settlement planning in Kinross-shire (Perth and Kinross Council).

In order to continue progress in the practical application of LCA, two fundamental points were seen as needing to be addressed.

First, there is a need for some descriptive guidance for planners to enable a greater understanding of the information contained in these documents. Not all Local Authorities, or Planning Departments, have their own landscape architects who could assist with the interpretation and application of the findings of the LCA. The need to raise awareness raising extends to the local political level, such as councillors, and there is also a need to raise awareness of landscape issues in relation to economic and other factors.

Secondly, there is a need to address how landscape issues fit with other considerations, and how important landscape is in relation to other determining issues such as socio-economic and community well-being, etc. LCA is clearly not a panacea, and there needs to be greater clarity regarding the role of the LCA in the planning process. The range of issues that arise in determining a development proposal, or in making a decision in relation to a Local or Structure Plan, are wide and varied. The LCA needs to be better integrated with these other considerations and therefore more effective in influencing the planning process.

18.4 Delivering the findings of the Landscape Character Assessment through the planning system

It is rare for decisions to go against the Development Plans, and in this sense they provide a strong framework in which to place the findings for the LCA. However, there is clearly a need to establish the relative importance of landscape and to develop techniques for effectively incorporating it into these important strategic documents. It takes a long time to establish new practice, and the LCA needs to be integrated properly into the Structure or Local Plan before the key findings of the LCA documents will be effectively adopted.

Planning Authorities are seen as being able to validate and create confidence in the findings of LCA, by incorporating them into the Development Plan and other land use strategic documents. Some LCAs provide a vision, new landscape frameworks or strategies which could be promoted and formalised through the Development Plan, although it is recognised that the Planning Authority alone could not deliver this vision.

Clarity is needed regarding the application and function of the LCA in the planning process, together with the various other approaches currently promoted, such as the National Heritage Resource Assessment (NHRA). The suggestion is that LCA may be getting lost amongst a multitude of other approaches and the many forms of advice to Planning Authorities.

18.5 Developing a more integrated approach to LCA

The majority of the existing LCAs, carried out as part of SNH's national programme, are seen as not including sufficient reference to local perception, or incorporating the perspectives of the communities living within the areas surveyed. Whilst it is recognised that not including the community perspective had been a weakness of much of the work carried out under SNH's national programme, the existing LCAs are supported and endorsed by the relevant Local Authority and therefore in a sense do reflect the public or the stakeholder. Although this is obviously only to a limited degree, they are nevertheless felt to be more than just a professional view. There is, however, clearly scope for this aspect to be improved upon and developed further. Examples were discussed of developing methodology for involving the community at a local level in the LCA, but they highlighted both the opportunities and difficulties to achieving this. Involvement of stakeholders and the public is recognised as being difficult, but clearly an area that should be covered. Mobilising a local community to take a landscape interest on would almost certainly increase the weighting given to landscape as an issue by councillors and decision makers. The importance of engaging landowners and people who can influence landscape change was also raised, and reference was made to the potential of the Environmental Capital Approach (Land Use Consultants, 1997) which seeks to build consensus among the many stakeholders.

LCAs have tended to recognise wider landscape issues, but not necessarily local issues, which can be equally important. Development of a more local approach, that includes promotion of local identity, could be advanced if LCA was seen as a part of community plans or the community planning process, and included in Rural Guidance and Rural Development Strategies. The spiritual and experiential aspects of landscape were also felt to be important and under-represented, and there is a need for a system or method which allows these to be better integrated into planning decisions.

18.6 Future directions for Landscape Character Assessment

Since the first guidance on Landscape Assessment (Countryside Commission, 1993) was produced, there has been a growth in the application of LCA. Many of the issues raised in the discussion session pointed towards the need for a review of the methodology of LCA and clearer descriptive guidance on its potential applications.

SNH and the Countryside Commission have commissioned a research project on LCAs to provide further guidance to help landscape policymakers and practitioners, particularly Local Authorities, to understand, conserve and enhance the character of the landscape. The guidance will describe and illustrate the methodology of LCA and provide examples of applications to land use, planning and management. The aim is to make the LCA an essential tool for the target audiences, especially policymakers and practitioners, but it is recognised this will only happen if there is a clear and definitive guide to the process and that this guide is well illustrated by applications to a broad range of land use planning and management situations. Now that the national programme is complete in Scotland it is time to seek to promote and explain LCA's use as a tool for landscape planning and management.

The evaluation side of the LCA clearly needs to be strengthened and the Environmental Capital Approach (Land Use Consultants, 1997) could perhaps be adapted to deliver a more integrated approach. The Environmental Capital Approach assesses the importance and benefits of features to relevant stakeholders, at various geographical scales. A pilot project in the North Pennines (Land Use Consultants *et al.*, 1999) raised two important findings; first that there is difficulty with the spiritual aspects of landscape and interpretation of landscape, and second that people find it difficult to separate community landscape and socio-economic issues. The professionals and agencies need to develop a more integrated approach, and identify and strengthen the link between these local issues, e.g. transport and the landscape.

There is now a gradual movement away from the Common Agricultural Policy into an era where the emphasis is being placed on more sensitive rural development and management. This creates an opportunity to take a more strategic approach in relation to addressing landscape issues. There is, however, some confusion regarding the priorities of Government and in knowing what weighting Planning Authorities are to give the various environmental directions from Government.

The Scottish Office is keen to promote good practice through National Planning Policy Guidelines and Planning Advice Notes, and, although this has long lead-in time, it is possible the LCA will increase and develop a role in Community Plans and rural guidance or policy. The Scottish Office recently launched *Towards a Development Strategy for Rural Scotland: the Framework* (Scottish Office, 1998), and will be piloting Path Finder Areas.

The aim will be to charge Local Authorities with developing community participation and through this develop a COMMUNITY PLAN. Perhaps this will be an important future opportunity to involve the community with the LCA process and incorporate community perception into the land use and development process.

References

Countryside Commission (1993). *Landscape Assessment Principles and Practice.* Cheltenham, Countryside Commission.

Land Use Consultants (1997). *Environmental Capital: a New Approach. Provisional Guide.* Unpublished report to the Countryside Commission, English Heritage, English Nature and the Environment Agency.

Land Use Consultants, Sheffield University Department of Landscape and CAG Consultants (1999). *An Integrated Approach to Environmental Planning and Management in the North Pennines; Environmental Capital Pilot Project.* Unpublished report to the Countryside Commission, English Heritage, English Nature and the Environment Agency.

Scottish Office (1998). *Towards a Development Strategy for Rural Scotland; The Framework.* Edinburgh, The Scottish Office.

19   CULTURAL ISSUES

T Chris Smout, Charles Withers, Lesley MacInnes and Fiona Lee

19.1 Genealogy and the historical dimension

The genealogy of landscape was outlined in a paper prepared by Charles Withers (Box 19.1). It is a geographical term reflecting ideology, power and politics through time, and it might be impossible to separate landscape from its cultural context. The development of landscape was traced from its early visual origins, its later use in scientific investigations as a material thing that can be recovered, uncovered or discovered, and importantly its development this century to include cultural meanings. Understanding landscape as a way of seeing, a point of view, recognises that landscape is a social product reflecting the views of those in authority. We do not all see landscape the same way, but if all landscapes are important, some may be deemed more important than others and, therefore, some ways of seeing landscapes are more important than others.

The relevance of the historical dimension to modern landscapes was stressed. Cultural landscape is traditionally taken account of as sites, but since landscape embodies traces of former field systems and other indications of past land use, it can legitimately cover extensive areas of landscape or connect blocks of landscape. Landscape character assessment (LCA) incorporates the built heritage in a restricted way, and does not recognise the full time-depth or the relationships between ancient and modern landscapes. Historic Land Use Assessment (HLA) (Dixon and Hingley, this volume) is undertaken at a larger scale and complements LCA. However, ideal landscape assessment should also address less tangible elements - like place names, artistic associations, history and mythology - that are important to people. These are not recognised in HLA. People's traditional land use, design and skills contribute to both the local distinctiveness and the sense of place. By identifying historical continuity and past changes, the human dimension of cultural diversity is added to what is seen as natural beauty. Landscape assessments should link the past, present and future, leading to more informed decisions about management options.

Because it is a palimpsest of past activity, all landscape is cultural landscape, often with immense time-depth. LCA marginalises that, and is more interested in natural processes. All landscape is culturally perceived. Because 'beauty is in the eye of the beholder' people care about place with a passion that has legitimacy but little scientific basis. People see different things in the same place: an empty Highland strath may seem lovely to the traveller, but a source of rage to others recalling the Clearances.

19.2 Language and values

The discussion focused on a number of issues. LCA is as objective as it can be, and hence has reduced the degree of subjectivity. The objective elements were associated with land, its description, and its susceptibility to GIS. The subjective elements were referred to as 'scape', the differing perception of different people and public preference. Subjectivity cannot be avoided, and complete objectivity is impossible, as judgements have to be made. However, both objectivity and subjectivity are needed, fitting the technocratic approach using trained knowledge with the critical approach that is aware of context and values. A participatory approach is needed, and evaluation should be separate from assessment.

Is nature conservation any more objective than landscape assessment? The question arises because we decide and differently 'value' the languages we use. The language of LCA is laden, and we personalise the landscape in using words such as 'character', 'identity', 'sensitivity' and 'tolerance'. What landscapes are we making now, and how would they be regarded in the future? It was suggested that Glen Lui in the eastern Cairngorms, where SNH's aim of restoring a Caledonian Forest has been modified by Historic Scotland's considerations, would be seen as a great conservation landscape of the late twentieth century, an artefact of quango power!

Returning to LCA, it was agreed that it must be done for a purpose, whether that purpose is objective or subjective. However, who determines the values? Who will be affected by change? LCA results in an intimate understanding of the landscape, and is a way to raise awareness and appreciation. LCA is also a tool to manage change, and to change it in the best way needs an understanding of its value, a rather circular argument! On the objective side, LCA is an inventory. Therefore, it is essential to keep description separate from perception, using both the professionals' expertise and the knowledge of local people. Putting a value on the landscape would follow. Interestingly, there was no agreement as to whether quality should be included in landscape assessment or follow later as part of a separate process (e.g. in developing strategies, or in considering designations).

19.3 Landscape change and mapping character

We cannot preserve everything, but it is useful to leave a mark that the future can look back at with pride. There is a perception that LCA seeks to maintain the *status quo*, fossilising a landscape that the consultants want to maintain, for example hedgerows that have not always been present. The opposing point of view is that we cannot maintain the *status quo* because landscape changes, but an assessment helps inform choices about future changes in the landscape. For example, the impact of farmers on the landscape needs to be recognised, and it also needs to be recognised that Government policy impacts on the landscape both in the agricultural support mechanisms and in the advice given to farmers. Agricultural landscapes are being reshaped by politics at he global, European, national and local levels.

Much of this is mappable, but how do we integrate language, stories or poetry into LCA, as they are not mappable? It is possible to incorporate 'local' or 'native' knowledge into a GIS, but difficulties arise in understanding the incommensurability between GIS (with its emphasis on measurement and overlay) and local meanings. Place names can help; for example, the Ordnance Survey *Original Object Name Books* were used in the 19th century. It was recognised that we must understand language to understand place, but yet the dominating science is object oriented.

19.4 Is SNH's LCA programme justified?

There was an interesting discussion of SNH's landscape character assessment programme, raising far more questions than answers. For example, is SNH going about it in the right way, and why did it not incorporate cultural landscape from the start? How will the LCA link with its programme on Natural Heritage Zones? It was felt that SNH needs information when taking decisions, hence it is important to document what is there and to know what is causing it to change. This is essentially the 'state-pressure-response' model that is being used by SNH in its programmes of audit. GIS is merely a tool, but it will not make decisions. LCA makes a data contribution in understanding the state of an important component of the natural heritage, and the pressures acting upon it. The challenge is to link scientific and non-scientific (or aesthetic) perspectives.

The LCA programme might have underplayed the cultural side, and it is therefore not totally complete. It has been worth doing and could be improved to incorporate more of the cultural factors of the landscape, both those in folklore and those is an archaeological sense. Therefore, SNH might need to think about how people could play a greater rôle in adding local meanings to the database that is landscape character assessment.

Box 19.1 Landscape: a partial genealogy of a geographical term

Like the terms 'nature' and 'culture' (for Raymond Williams (1963), amongst 'the most complicated words in the English language'), 'landscape' enjoys no single simple definition. Several meanings have been identified and are employed in geographical enquiry. Whilst the claim has been made that the term was understood in medieval England to connote the land controlled by a lord, the term is usually traced to the early 17th century, to the influence of Dutch and Italian art, and, through the idea of *landskip* or *landschap*, to mean the appearance of an area, more particularly the representation of scenery. Landscape thus has intellectual 'origins' as a visual term (and is understood as such in landscape art of the 18th and 19th centuries). The term is also used within the physical earth sciences in ways which see landscape as physical environment, a sense which owes its origin to the empirical investigation of Nature and, in geography, to links to the *Landschaftkinde* interests of later 19th century German geography.

The idea of a cultural landscape appears in the 1920s, notably in France, Britain and the USA, in which sense landscape was understood as 'an area made up of a distinct association of forms, both physical and cultural'. Amongst the central concerns of geography at this time was the question of tracing the change from 'natural' landscape to 'cultural' landscape, a question which focused on culture as an absolute and shared thing, and upon the 'morphology' of landscape. From the 1950s, it is possible to determine a clear focus in landscape as historical artefact, a material 'thing' capable of being 'recovered'. This tradition is apparent in landscape history, landscape studies and finds particular expression in, for example, Hoskins (1955) and Muir (1998). Since the 1980s, attention has turned to the idea of landscape as a way of seeing, an area of enquiry in human geography that stresses the ideological bases to landscape, draws methodological insight from art history and places landscape in relation to social and political systems. Landscape in these ways is not alone a material object and is always implicated in questions of power, authority and perspective (Duncan, 1994).

Landscape as culture, a way of seeing: why this might matter in considering 'landscape character'

In this understanding, in which particular attention has been paid to the iconography of landscape, landscape is understood in relation to the ways in which a particular class or

---

other (dominant) group in society has represented and constituted its own views. There are important connections with the visual 'origins' of the term here, notably in the terms 'perspective', 'prospect' and 'point of view': i.e. that a landscape is always visualized from somebody's point of view, that it encompasses a particular perspective and that it affords a prospect in two related senses, outwards in space (outwards, that is, from one's point of view), and forwards in time (see Cosgrove, 1984, 1985; Cosgrove and Daniels, 1984). In important ways, this emphasis within human geography on landscape as culture, upon the iconographic bases to national identify, for example, and upon the ways in which landscape as a visual term is rooted in questions of power and authority, has close connections with the 'original' meaning of the term. And, for a variety of reasons, this approach has a part to play in considerations of what is understood by 'landscape character'. Thus, consideration of landscape as a way of seeing

- allows us to recognise that whilst empirical scientific methodology accords a primacy to observation we do not all see the same things;
- moves beyond the reading of landscape as alone being a material natural object capable of having attributes, 'personality' or 'character' accorded it without reference to the social and political constitution of those terms;
- stresses the need to examine the *particular* senses in which the term is used in relation, for example, to matters of gender, local meaning, social and cultural difference;
- places matters of power and social authority (and, by implication, institutional position) centrally within any explanation of the term 'landscape';
- ensures an understanding of the complex historical genealogy of the term 'landscape' as a matter of contemporary significance; and
- should caution against the uncritical equation of 'landscape' with 'nature', 'scenery', 'environment' or 'countryside'.

References

Cosgrove, D. (1984). *Social Formation and Symbolic Landscape.* London, Croom Helm.

Cosgrove, D. (1985). Prospect, perspective and the evolution of the landscape idea. *Transactions of the Institute of British Geographers*, **10**, 45-62.

Cosgrove, D. and Daniels, S. (Eds.) (1984). *The Iconography of Landscape.* Cambridge, Cambridge University Press.

Duncan, J. (1994). Landscape. In Johnston, R., Gregory, D. and Smith, D. (Eds.), *Dictionary of Human Geography.* Blackwells, Oxford, 316-317.

Hoskins, W.G. (1955). *The Making of the English Landscape.* London, Hodder & Stoughton.

Muir, M. (1998). Geography and the history of landscape. *The Geographical Journal*, **164**, 148-154.

Williams, R. (1963). *Culture and Society, 1780-1950.* Harmondsworth, Penguin.

20 MONITORING AND MEASURING LANDSCAPE CHANGE

R.V. Birnie, E.C. Mackey, S. Leadbeater and A. Bennett

20.1 Introduction

The aim of this session was to address two questions: what is missing from current efforts to monitor and measure landscape change, and, leading from this, what is needed in the future? This chapter is structured around these two questions. The views expressed reflect the range of perspectives aired by the discussion group and are not necessarily either those of the authors or of all the individuals in the group.

20.2 What is Missing?

There is a wide range of statutory and non-statutory programmes for monitoring change in the countryside in the UK. The former include the various Agricultural Censuses, the latter such initiatives as the National Countryside Monitoring Scheme (Mackey *et al.*, 1998) and the Countryside Survey sponsored by the Department of the Environment, Transport and the Regions. Most of these programmes are identified in Table 20.1 and they provided the background to the discussion.

Table 20.1. Survey and monitoring programmes identified by Birnie *et al.* (1995)

Statutory Land Use Surveys

> June Agricultural Census
> Integrated Administration and Control System (IACS)
> Monitoring Agriculture with Remote Sensing (MARS)
> Forestry Authority Woodland Inventory

Non-Statutory Land Use Surveys

> Baseline surveys
> • Ordnance Survey Land-Line.93
> • Department of the Environment, Transport and the Regions/Institute of Terrestrial Ecology Land Cover Map of Great Britain
> • Scottish Office/Macaulay Land Use Research Institute Land Cover of Scotland 1988 (LCS88)
>
> National Monitoring
> • Department of the Environment/Institute of Terrestrial Ecology Countryside Survey 1990 and related CORINE mapping
> • Scottish Natural Heritage National Countryside Monitoring Scheme (NCMS)
> • Department of the Environment Land Use Change Statistics
>
> Regional Monitoring Initiatives
> • Countryside Commission Monitoring Landscape Change in the National Parks
> • Central Valley and Cairngorm Look Back Studies
> • Environmentally Sensitive Areas Monitoring
> • National Scenic Areas Monitoring

There are two fundamental issues that need to be addressed. Firstly, 'why are we so concerned with change?' and secondly, if change is inevitable, 'why should we monitor it?'. The first question relates to fundamental aspects of the human condition and is essentially a psychological one. Recent work using personal construct psychology applied to landscape evaluation has highlighted the critical importance of change, particularly the form of change, in determining our valuation of landscape (Harvey, 1995). The importance we attach to change therefore may be linked to the importance change has in shaping our views of how much we value something. The second question reflects a view that we are not capable of directing change – it is 'inevitable'. Monitoring change, therefore, suggests a commitment to intervene in the processes of change, either directly or indirectly.

If we accept the view that landscape monitoring is of value only where it is linked to action (i.e. it is 'surveillance with a purpose'), and that action has clear objectives, then it can be assumed that it is not possible to define a monitoring programme without a set of objectives. This would require a policy framework, which would set out these objectives. The group was concerned that there is no articulated policy framework which sets objectives for landscape monitoring in the UK.

Turning to one of the major relevant initiatives, Landscape Character Assessment (LCA) developed by SNH, the group asked the question, given the lack of actual landscape monitoring initiatives to date, 'could the LCA be used in landscape monitoring?'. It is believed that the LCA is primarily an exercise in landscape classification and has not been developed with a monitoring purpose in mind. This raises a number of critical issues such as its subjectivity, its repeatability and the explicitness or otherwise of the valuation system that underlies it. The group was uncertain about whether the LCA provided an adequate baseline for monitoring landscape change in Scotland and felt this needed to be more critically assessed.

Because 'landscape' is a multi-dimensional concept, a critical element of our discussion revolved around 'what needs to be monitored/measured?'. It was pointed out that present monitoring efforts focus on those elements of the landscape that lend themselves to measurement (e.g. Countryside Surveys focus on plant species and land cover). The more intangible elements including cultural ones (e.g. 'sense of place') are seldom included, neither are the processes of change that change the relationships between people and landscape. The discussion revolved around the issues of whether there are indicators that might adequately capture some of these intangible elements. If not, then 'only the things that are measurable become important'. The group was unable to define a set of candidate indicators to show how demographic, technological and cultural changes, which affect landscape change, might be developed.

The final strand of this part of our discussion examined the question of 'how to monitor?'. This picked up the earlier discussion on the multi-dimensional nature of 'landscape' and concluded that no single monitoring approach would be appropriate. It was most likely that different monitoring methods would be required for the different dimensions of landscape change. A critical point is that monitoring should cover not only what is changing, but also seek to embrace measures of why it is changing: the concept of drivers of change. A specific point in this regard was the monitoring of changes in valuations of landscape (e.g. through use of techniques such as cost benefit analysis or contingent valuation) as related to different user groups. This means that we do not seek to impose our values through the monitoring system, but to reveal other people's values and how they are changing. The critical conclusion is that we need to create a 'living, breathing model' which can be assessed, used and applied by society as a whole, and not just a 'dusty report'.

20.3 What is needed?

The group identified the following four key issues that needed to be addressed.

- Landscape monitoring has to be set in the context of UK and Scottish policy on sustainable development. There is a need for some clear policy objectives, which can be translated into a monitoring protocol. There was a general concern that the relevant policy arenas (agriculture, natural heritage, environment, etc) had to be viewed together, in a more integrated way, to provide a more 'joined-up' approach.
- Current 'landscape' monitoring initiatives reflect the sectoral nature of rural policy and lack a coherent vision, which would enable them to be better integrated. There is a general fascination with 'effects' rather than 'causes' in monitoring, and a general avoidance of the less tangible aspects of landscape that might actually be more important culturally.
- The present 'user' community misses some of the major landscape consumers, in particular the tourist industry. Some effort must be made to include these other groups in the debate about landscape change in Scotland.
- Access to data is a critical issue. This is seen as a critical limitation to the evolution of new ways of informing society and shaping future change in the light of existing information and available knowledge.

References

Birnie, R.V., Morgan, R.J, Bateman, D., McGregor, M.J., Potter, C., Shucksmith, D.M., Thompson, T.R.E. and Webster, J.P.G. (1995). *Review of Land Use Research in the UK.* Unpublished Macaulay Land Use Research Institute report.

Harvey, R. (1995). *Eliciting and Mapping the attributes of Landscape Perception: an Integration of Personal Construct Theory (PCT) with Geographic Information Systems (GIS).* PhD Thesis, Heriot-Watt University.

Mackey, E.C., Shewry, M.C. and Tudor, G.J. (1998). *Land Cover Change: Scotland from the 1940s to the 1980s.* The Stationery Office, Edinburgh.

21 TECHNOLOGICAL DEVELOPMENTS IN TOPOGRAPHIC DATA CAPTURE, INTEGRATION, MANIPULATION AND MANAGEMENT

Marshall Fairbairn and Alistair Law

21.1 Introduction

This workshop was introduced with an overview of how Ordnance Survey, the National Mapping Agency, captured, maintained and makes available the topographic database for Great Britain. It then focused on the following issues

- national supply, availability and cost of a range of digital mapping datasets,
- adherence to consistent national standards,
- National Interest Mapping Service Agreement (NIMSA) investment to provide enhanced, up to date, quality products,
- working with partners and collaborators to provide value added products,
- service level agreements and the development of closer relationships with customers, and
- exploiting new technologies for the common good.

21.2 Dataset availability

A full range of digital map datasets of Great Britain is now available to customers to enable them to implement and develop their geographic information systems (GIS). These products cover a number of 'themes', including

- a topographic theme, with various scales of datasets both in vector and raster form, such as large scale Land-Line to the small scale BaseData.GB;
- a landform theme, with height datasets such as Profile and Panorama both as contours and as digital terrain models;
- a boundaries theme, with voting and administrative boundaries such as Boundary-Line and ED-LINE; and
- a gazetteer theme, incorporating address and road information from ADDRESS-POINT and OSCAR.

The availability of the national coverage of data, reduced costs, increased power and widespread use of computer technology were accepted as the drivers for the increasing awareness and pace of development of GIS across the business community. The needs of the maturing GIS users constantly push suppliers to provide data that is fit for purpose with costs that can be justified as bringing benefits to the business.

Concern at the cost of obtaining data was raised particularly by the education community who, with limited funding, found it especially difficult to use OS data in the course of their teaching. The meeting was advised of the ongoing negotiations between Ordnance Survey and JISC to establish an agreement where a package of digital mapping products would be made available at an affordable cost. Information was also presented on the pricing policy review underway within OS where the aim is to introduce a simplified costing structure including supply, licence usage and hardcopy output.

To counteract the perceived high cost of obtaining data, OS recommended the formation of cost effective, customised service level agreements (SLAs) between OS and a number of similar small enterprises as the best route to acquire and use OS data. The SLA established with the Local Authorities demonstrated that such a non-standard deal provided the impetus to kick start their use of GIS. Since setting up this agreement OS has encouraged and entered into many similar arrangements to supply a package of data products, including supply, maintenance, and licence activities for a fixed period at an acceptable and justifiable cost.

21.3 Management of data

Standards for geographic information are increasingly being recognised as important if the benefits of GIS are to be realised. Ordnance Survey's role in new government initiatives aimed at providing access to the mass of government, local authority, utility and commercial data was examined and discussed. It was accepted that to provide reliable up to date and consistent information at a national level required the application of pre-defined, agreed and quality assured standards. Examples of this commitment were the successful pilot trial in Bristol of the National Land Information System (NLIS) whereby datasets from a variety of diverse sources are linked to OS map backgrounds to provide a comprehensive report on land and property available for sale. Ordnance Survey have finalised the contract with the local authority community to compile and publish via the Internet a National Street Gazetteer compliant with Part 1 of BS7666 (Anon., 1993/94). It is also working closely with the Department Of Environment Transport and Regions, Local Authorities and English Partnerships to create a National Land Use Database for England and Wales.

The National Interest Mapping Service Agreement (NIMSA), signed on 13 October 1998, enables OS to deliver services to support the national interest. The major components of this agreement are

- uncommercial mapping activities,
- national consistency of content, currency and style, and
- the underpinning infrastructure of mapping.

The activities that have been identified as requiring full or partial funding fall under two broad categories, i.e. those that are directly connected to maintaining the National Topographic Database (NTD) and those related to the provision of the technical, administrative and advisory duties as the national mapping agency.

This commitment provides funding to OS to pursue activities directly connected with maintaining and investing in the NTD. Prime amongst the listed NIMSA activities is the programme nationally to revise and update rural, mountain and moorland mapping by October 2000. More up-to-date mapping allied to a programme of positional accuracy improvements for some 1:2,500 scale mapping, and the proposed conversion of the large scale database to a topologically-based data specification, will enable the development of a new geospatial environment management system, needed to produce richer, more explicit defined data for the benefit of all customers.

21.4 Looking to the future

Other activities identified to provide the mapping infrastructure necessary to meet the needs of modern, proficient and demanding customers are the creation of a new 'Landplan' 1:10,000 scale map series derived from the large scale database, the creation of a new 1:25,000 scale database, the replacement of existing Pathfinder maps with Outdoor Leisure Maps, and the preparation of a new Explorer series of maps.

To satisfy the demand for access to horizontal and vertical mapping datums by users of Global Positioning Systems (GPS) a modern reference frame for position fixing will be created, known as the National GPS Network. Benefits to users will include easy access to nationally positioned facilities, reduced costs and a maintained system.

As a national mapping agency it is essential that an all hours capability is provided to react to requests for delivery of accurate up to date maps to meet any eventuality, from a local incident to a major national emergency. It is also recognised that OS must be in a position to represent the British Government at national and international forums and also to respond to the needs of the educational sector in the field of geographic information.

Rapid advances in technology offer opportunities for OS to improve its business efficiency and effectiveness, provide a greater range of affordable products and improve its services to its customers. New data capture methods such as the employment of hand held computers, the Portable Revision Interactive Survey Module (PRISM), by the 450 field surveyors enables direct digital capture of topographic detail with transmission direct to the database in Southampton. Aerial photography plays an important role in meeting the revision targets for the large scale database. Over 65,000 highly detailed photographs are taken each year to provide information on changes to the landscape, such as industrial development, major roads, forestry or changes to the coastline caused by erosion or variations in tide patterns. This same photography can also be used to create three dimensional models of the landscape, offering planners and engineers the ability and the tools to view and analyse virtual worlds.

Internet technology, with its ability to provide instant and widespread access to information, offers many opportunities to data providers such as OS. Access to on-line geographic information opens the doors to a vast range of data; however, this must be managed and controlled to ensure the issues of intellectual property rights and copyright charges are properly accounted for.

21.5 Conclusions

Through initiatives such as 'Joined-up Government' and 'Open Government' the programme has been set for government to move into the information age so as to provide improved services, at lower costs and to increase public trust in government To meet this challenge of increasing demand for immediate access to all forms of public information, government aims, through its 'Invest to Save' budget, to provide the significant resources necessary to deliver modern, professional and efficient services.

Ordnance Survey, through the agreed use of NIMSA funding and the proposed move to

Trading Fund Status, also plans to meet this demand for new investment. Through closer relationships with customers, consultative committees and from market research, OS aims to establish what geospatial products and services it must develop to meet future demand.

To encourage widespread use of geospatial data OS will continue to seek to provide a full range of up-to-date data products fit for the modern user at an affordable price. Key to this aim investment will be made to exploit advances in technology to collect, process and deliver these products by methods designed to drive down costs. OS will also embrace other challenges that will impact upon its business, such as changes in legislation. Overall, Britain's National Mapping Agency is committed to

- develop products and services to the quality and standards that a modern advanced country demands and needs;
- develop stronger partnerships with data providers in the public and private sectors;
- engender open and fair competition with all customers, and
- exploit the continuing wider availability and everyday use of information technology.

References

Anonymous (1993/94). *BS7666 Spatial data-sets for geographical referencing. Parts 1-4.* London, British Standards Institution.

PART FIVE

A CONCLUDING PERSPECTIVE

22 GEOGRAPHIC INFORMATION FOR RESEARCH AND POLICY: A NORWEGIAN LANDSCAPE PERSPECTIVE

G. Fry, O. Puschmann and W. Dramstad

Summary

1. Landscape classification and assessment methods are currently attracting interest from research and management. This interest reflects the limitations of administrative boundaries as land management units.
2. We need a critical assessment of local and national landscape mapping initiatives in order to avoid duplication and promote their wider use. The Norwegian landscape mapping scheme is described as a case study with similarities and contrasts to schemes used in the UK. Landscape character assessments offer one of the best chances for achieving an integrated landscape assessment and management.
3. There is unlikely to be one best approach to landscape assessment; existing methods therefore require testing and validating against clear aims. For example, their ability to identify which landscapes are vulnerable to which pressures and suitable for which management solutions needs to be assessed.
4. The boundaries of landscape regions should be flexible to reflect the dynamic nature of landscapes and the reasons for mapping them.

22.1 Introduction

In this chapter, the use of geographical information is examined for research and planning from a landscape perspective and with special emphasis on landscape character mapping. Since the 1960s, the use of digital geographic information in natural resource management has exploded in quantity and diversity. Primarily, this has been a result of advances in technology and their adoption by the landscape professions (Ripple, 1987; Levine and Landis, 1989; Haines-Young *et al.*, 1993; Wickham and Norton, 1994; Wadsworth and Treweek, 1999). The technological advances in managing geographic data, particularly in the field of geographical information systems (GIS), have been significant. Yet, the impact of GIS reflects not only technological development - the basic procedures were available 20 or more years ago - but also GIS availability and its increasing acceptance in natural resource management (Herpsager, 1994). Parallel to this development there has been a rapid increase in the availability of digital data. The essential issue here is the accessibility of data, since this has been a larger problem than whether data exists or not. Problems accessing map data (often collected with public funds) relate to ownership and user rights, confidentiality, lack of standardisation of data formats and the privatisation of mapping agencies resulting in a need to generate external income. Nevertheless, an increasing range of environmental data is available in digital map format. In addition, thematic data useful for planners, including cultural heritage and amenity resources, are also becoming available in digital map format.

These technical advances have led to a myriad of research and development projects whose wide-ranging approaches offer an extremely valuable set of case studies and trials of mapping and assessment methodology. One urgent requirement, however, is that these local and national initiatives are reviewed and critically assessed for wider use. There are many parallel and overlapping schemes under development but too little co-operation and co-ordination. The risk, as in all rapidly developing fields, is that we waste time 're-inventing the wheel'. New landscape mapping and assessment methods tend to become identified with individual institutes or people. As a result, they often end up competing with each other, rather than contributing to an on-going development taking the best ideas from each so as to achieve specific goals (see Lawton (1991) for a discussion of a similar problem in ecology).

Although most changes in the countryside are outwith formal planning procedures, the development of the rural landscape is increasingly planned through the application of agricultural and forestry grant schemes, and environmental regulations. The result has been a significant increase in the control of rural development in Europe both in terms of the area of countryside and range of developments under some form of local or strategic planning (Gjølberg, 1995).

22.2 Why a landscape perspective?

Advances in subjects with close links to planning have contributed to the theoretical basis for supporting landscape as an integrating concept. Landscape ecology, for instance, has forced us to consider the spatial structures of landscapes and how these affect the distribution and survival of species as well as the use and appreciation of the countryside by people (Forman and Godron, 1986; Selman, 1993; Herpsager, 1994; Forman, 1995; Fry and Sarlov-Herlin, 1995; Dramstad *et al.*, 1996). As planning units, landscapes are important as they often form the arena where issues are debated and conflicts resolved at a human scale. They are the perceptual units used by people to define the boundaries of current and historical cultural environments. Landscapes are integral components of national and local cultural identity and increasingly accepted as worthy of conservation (Hitier, 1998; Wascher, 1999).

There is a growing demand from politicians for a more integrated rural assessment, one that includes cultural and natural heritage components within a living countryside (Macinnes and Wickham-Jones, 1992). Landscape units offer a sound basis for the combination of natural and cultural interests in urban and rural planning (Nassauer, 1992). Landscapes also offer a focus for conservation in the wider countryside, expanding our perspective from sites to whole landscapes. Such trends in countryside management have stimulated the development of landscape mapping and assessment schemes. The former aim at description, classification and mapping, whereas the latter more explicitly aim at describing and assessing landscapes in order to capture human responses in addition to biophysical descriptions. In use, the two terms overlap since there is no mapping without value judgements and assessments often start with a description of the land form (Fry, 1998).

Drawbacks with the use of administrative boundaries in natural resource management have also stimulated the search for new approaches (Mitchell, 1989; Blankson and Green, 1991; Countryside Commission, 1993; Stanners and Bourdeau, 1995; Brabyn, 1996;). Administrative boundaries may poorly reflect natural units. For nature conservation, the areas of search used in both site evaluation and in setting targets are usually administrative boundaries although these may have little relevance to the distribution and hence management of species, communities, and habitats. Similar arguments apply to cultural heritage management where historical landscapes often cross administrative boundaries. Although we are still far from solving all the practical problems in co-ordinating the management of resources across administrative boundaries, the concept of using natural boundaries has caught on and is gaining wide support (English Nature, 1993; Wascher, 1999).

As well as providing appropriate spatial units for the management of natural and cultural resources, landscape character assessment provides a good platform for integrating different landscape interests. This is especially urgent since the sectoral basis for resource management remains a major source of countryside conflict and an obstacle to multiple-use planning (Fagence, 1988; Hertig, 1993). Management units that combine natural and cultural values pave the way for clearer and more comprehensive assessments of the value of areas and in developing coherent plans for future development (Wagstaff, 1987). Developmental work on the concept of environmental capital in the UK (CAG Consultants and Land Use Consultants, 1997) provides an example of comprehensive assessment.

22.3 The need for landscape classification

An overview of landscape resources requires a landscape classification. Management of landscapes without a classification system is analogous to developing a nature conservation strategy without a consistent system for classifying species or habitats. As landscape takes a more prominent place in land conservation strategies, it becomes more urgent to devise methods of landscape classification and assessment to target management measures and monitor change (Blankson and Green, 1991; Stiles, 1996).

For a landscape-centred approach to managing natural resources, the first step is to identify mappable units that are more relevant to resource management than administrative boundaries. The difficulty in selecting appropriate landscape units arises from the way natural and cultural resources react differently to the natural variation found across large geographic areas (Aldenderfer and Mascher, 1996; Host *et al.*, 1996; Wright *et al.*, 1998). Yet, for many natural and cultural resources there exist well-documented large-scale responses to variation in topography and climate (Allen *et al.*, 1990; Blankson and Green, 1991). Vegetation follows these trends, as do many animal species (Austin and Myers, 1995; Carey *et al.*, 1995). Not surprisingly most landscape classification systems are essentially biogeographical at a large scale.

At finer scales, approaches diversify through their weighting of different landscape parameters. The main differences are between methods with a strict focus on biophysical parameters (Austin and Myers, 1995; Carey *et al.*, 1995; Host *et al.*, 1996) and those which also include cultural and visual aspects (Countryside Commission 1993; Environmental Resources Limited, 1995; Stiles, 1996; Williams and Patterson, 1996; Wascher, 1999). There is no deep theoretical division between these approaches but significant differences in the choice of data themes used to identify landscape units. There is no right and no wrong approach without reference to the objectives for specific landscape classifications. This implies that there will never be a perfect method fulfilling the wishes of all potential users. Instead, we must continue to develop and judge methods according to their aims and utility.

22.4 Bottom up or top down landscape classification?

The dominant approaches to landscape classification use hierarchical divisive strategies that divide and sub-divide countries or larger geographic areas progressively into smaller and smaller units with uniform characteristics (Stanners and Bourdeau 1995; Wascher 1999). Variation within this approach reflects differences in the selection of landscape characteristics used in the analysis. Commonly, a mixture of geology, topography, vegetation and land use are used (Wascher, 1999). Analysis, whether by numerical methods such as ordination or classification (Gauch, 1982) or expert evaluation, results in a number of classes comprising units of similar characteristics which may occur in several geographic locations. These units may be further divided into smaller landscape zones to take more account of human influence and human perception of landscapes at finer scales. Alternatively, further sub-divisions may continue, being based on environmental parameters (see Mitchell (1989) for an historical overview).

Bottom-up approaches start with detailed studies of small geographic units and agglomerate these into increasingly larger ones. The disadvantages of this strategy are that it fails both to provide an overview at the start of the process and to recognise that geographically distant units may be very closely related, e.g. reflecting similar basic conditions and historical development. Some schemes use a combination of both approaches, working with detailed descriptions of landscape zones within already defined larger regional units, e.g. the Scottish Natural Heritage approach to landscape character assessment (Hughes and Buchan, this volume; Environmental Resources Limited, 1995). Bottom-up approaches have the advantage that they often include a significant involvement of local people. In addition, approaches starting at the local scale will give weight to important local landscape features which might be missed by top-down approaches (Tilley, 1994; Endter-Wada *et al.*, 1998). The symbolic significance of such features may be so great that they become important at higher levels of landscape classification or to the conceptual development of assessment methods. Hierarchical systems starting at the regional or larger scales, therefore, seem to have the greater advantage but special local features should be allowed to influence assessment or to define sub-units.

22.5 Objectively - subjectively derived landscape units

At fine scales (1-10 km), landscape mapping approaches more clearly divide into methods based on the numerical analysis of environmental data (Brabyn, 1996), and those which rely on expert or consensus paradigms to define landscape units (Zube, 1984). The former seek to find similarities between landscape types based on the prevailing natural conditions and hence similar biotopes or land use suitability classes. The latter openly address the importance of human use and perception of landscape in defining landscape character (see Table 22.1). There is, as yet, no widely-accepted methods of using environmental data to classify human responses to landscape numerically. Yet, groups of landscape professionals can assess landscapes in ways that involve natural and cultural heritage aspects as well as

Table 22.1. Arguments associated with objective and subjective landscape assessments. Recent approaches combine the best of both approaches to provide factual descriptions at different scale levels as well as to comment on the overall impression given by the various components of landscape. There is a clear trend to move from objective to combined methods as one shifts from broader to finer scale landscape classes.

Subjective methods

- capture 'sense of place' which numbers and description alone fail to do

- identify unique combinations of features which would be otherwise overlooked

- assess aspects of cultural identity, including weighting of culturally important landscape features (battlefields, home of author or poet, historical continuity, etc.)

- may include invisible aspects of landscape in assessments (especially the time dimension important in historical landscapes)

- give weighting to small but important features

Objective approaches

- communicate well between sectors since the data on which assessment is based is 'hard' and easy to identify

- facilitate comparisons between regions, as the methods are standard

- make assessments easier to defend as the results are repeatable

- allow data to be re-analysed for different purposes or as additional data layers become available

- make assessments more democratic as they are transparent

- allow map units to be re-classified as values or theory evolve

experiential responses. Advances in methodology, particularly standardisation of assessment procedures, add to the reliability of the approach and its wider acceptance.

Methods rarely, if ever, find themselves at one end of the scale between objective and subjective, but somewhere in the middle. Decisions about which landscape characteristics to use in a landscape analysis involve value judgements that give weighting to the chosen characteristics. This is impossible to avoid, but legitimate if made explicit and linked to clear goals. In the future, there is likely to be greater flexibility in methodologies (Aldenderfer and Maschner, 1996; Wadsworth and Treweek, 1999), better compatibility between experiential responses and GIS analysis (Nelson and Serafin, 1992; Llobera, 1996), and landscape professionals living more comfortably with mixed quantitative and qualitative approaches.

Most landscape mapping approaches have many similarities at larger scales (e.g. national regions). They reflect large-scale biophysical regions rather well and often the geographic regions familiar in school textbooks of 20 or more years ago. It is when moving from scales of hundreds to tens of kilometres that greater variation in approaches becomes discernible.

22.6 The Norwegian Landscape Mapping System

The Norwegian landscape mapping scheme offers a case study with both parallels and contrasts with schemes used in the UK. The system has several interesting features; it is hierarchical, flexible, user-based and continuously evolving. The method also has advantages associated with being developed by a central institute, the Norwegian Institute for Land Inventory (NIJOS). These include working with relevant national datasets, a large mapping infrastructure and close contact with landscape management needs (agriculture, forestry, environment, tourism, development, etc.) at national and local levels.

22.6.1 *Historical Background*

Based on a growing dissatisfaction with administrative boundaries as planning units in the 1980s, several approaches to landscape-based planning were developed in the Nordic countries (Nordisk Ministerråd, 1987). The Norwegian contribution was based primarily on the USDA system of scenic planning (USDA Forest Service, 1974) and adapted to Norwegian conditions by Bruun (1987). The project criticised administrative boundaries, which often cut across continuous landscape units or included landscape types of widely differing qualities. Landscape regions were proposed as an alternative, providing a better framework in which to document landscape change, assess its possible consequences and develop new integrated planning strategies.

During the 1990s NIJOS further developed and tested the methodology to provide new management and planning tools, especially in relation to the rapidly developing, landscape-based tourist industry. A national reference system for landscape assessment has been produced with support from the Norwegian Ministries of Agriculture, Environment and Industry (Puschmann, 1998).

22.6.2 *Purpose of the Norwegian landscape mapping system*

The main purpose of the national reference system for landscapes is to increase understanding of landscape as a resource in planning and management. It also supports the Norwegian focus on development based on natural resources (Kamfjord *et al.*, 1997). A further aim is to indicate the location and extent of environmental problems linked to specific landscape types (see example in Figure 22.1). Such information will be an important aid in the application of financial or control measures to ameliorate environmental problems.

The resultant maps of regions provide planning frameworks at different spatial scales, from national strategies to local plans. The pressure on rural and urban areas in Norway is increasing, accompanied by larger numbers of stakeholders involved in land use issues. Landscape assessment is one way of providing a common information base for the diverse actors in land use debates.

22.6.3 *Method development for Norwegian landscape mapping*

The Norwegian method aims to provide a hierarchical classification of landscapes at three spatial scales; the national level (landscape regions), the regional/county level (landscape sub-regions) and the municipality level (landscape areas). At the national level, there are 45 landscape regions which are divided into 444 sub-regions (Elgersma, 1996). The classification of units is based on a systematic description of six landscape components; landform, geology, water, vegetation, agriculture (land use), and settlement/infrastructure (see Figure 22.2). Interactions between the different components form the basis for describing 'landscape character'. Geographical units which share common features are grouped into landscape regions, sub-regions or landscape areas according to the geographical scale of interest. The national regions are based on large-scale biogeographic

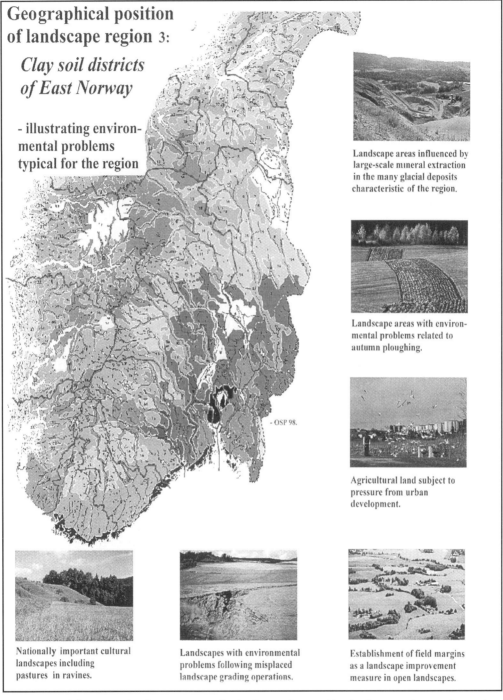

Geographical position of landscape region 3:

Clay soil districts of East Norway

- illustrating environmental problems typical for the region

- OSP 98.

Landscape areas influenced by large-scale mineral extraction in the many glacial deposits characteristic of the region.

Landscape areas with environmental problems related to autumn ploughing.

Agricultural land subject to pressure from urban development.

Nationally important cultural landscapes including pastures in ravines.

Landscapes with environmental problems following misplaced landscape grading operations.

Establishment of field margins as a landscape improvement measure in open landscapes.

Figure 22.1   The national reference system for landscape data enables landscape regions to be used as geographical units for analysing environmental problems and development potential.

LANDSCAPE COMPONENTS

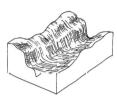

Main landform
Major topographic units.

Geological details
Geology and geomorphology, minor landforms.

Lakes, rivers, fjords and coast
Size and form. Streams, white water, and waterfalls. River valleys, lakes, beaches, coastline.

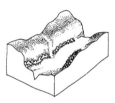

Vegetation patterns
Main vegetation communities, cultural influence, patterns, structure and mosaics.

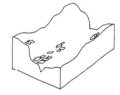

Agriculture / land use
Pattern, size and structure of agricultural units. Land use, forms of production, diversity rotations, mono-cultures etc.

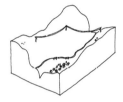

Settlement & infrastructure
Traditional buildings, harmony / disharmony between old and new. Roads, power lines, industry. Historical continuity

(Ill: Gaute Sønstebø, NIJOS 1993)

Landscape character
The character of a landscape is the result of its landscape components and the total impression they make.

Figure 22.2 Landscape character is the result of landscape components and the total impression they give. Each landscape component is studied through compilations of geographic data plus site descriptions. The character of the landscape is an essential feature of the classification and mapping of landscape regions.

characteristics, which necessitate a degree of simplification and generalisation limiting their value for detailed planning. Sub-regions, therefore, provide the framework of reference.

Borders between landscape regions reflect changes in dominant landscape attributes. At the national level, landform will often be the determining factor. Landform is also used at the sub-region level, but at this finer scale they reflect repeating formations, e.g. drainage catchments. If a part of a region includes a significant new element or differs in any way, a new sub-region results. When landform is less significant to the visual character of a region, the other components of landscape increase in importance. At the local level, sub-regions are divided into landscape areas, based on character determined by the six landscape components and the interactions between them. At this level, the aim is to develop a tool useful in detailed planning at the municipal level, and to visualize local variation and identity.

22.6.4 *Current uses of the Norwegian landscape mapping scheme*

Progress of the national mapping scheme includes a full systematic description of landscape character for the 45 landscape regions and for 60 of the 444 sub-regions. At the local level, however, only a small sample of landscape areas has been fully described. Municipalities can be completed by request, as is often the case in relation to development projects or new strategic plans. Full descriptions of all 444 sub-regions are planned for completion by the year 2002.

Landscape mapping has received a positive response from planning authorities, as evidenced by requests for detailed landscape assessments. In addition, Lofoten, a well-known tourist area off the west coast of Norway, provided a case study for testing the value of sub-regional landscape descriptions for planning tourist development. The case study identified methodological strengths and weaknesses for further development, but also demonstrated the benefits of landscape assessment as a tool for identifying natural assets (Kamfjord *et al.*, 1997). Locating landscapes with the characteristics tourists identified with the Lofoten region was one step in this process. These were special coastal landscapes with a combination of traditional fishermen's cottages (rorbu) and views of the midnight sun in summer. Although the image is marketed all over the world, very few places were found to offer this combination of assets. Identifying their geographical location assisted in formulating local development plans.

22.6.5 *Validation and future development*

The Norwegian system is aimed at supplying the needs of planners and developers. Case studies provide concrete examples of use, while the method is constantly undergoing further development and validation. When describing the history of the Norwegian system, the following quote is relevant to the ongoing debate in landscape mapping.

> " *We should, however, accept that landscape regions are not to be defined once and forever within exact boundaries. They will need to be modified and corrected, as additional information becomes available. Regionalization is a part of the rotating process of landscape planning, and can, as such, never be brought to a final conclusion.* " (Nordisk Ministerråd, 1987, p. 19).

An exciting recent development is the way landscape maps have been linked to national and local databases (Puschmann, 1998). This makes it possible to examine the type and production of farming systems, housing, population, industry and other geographic information within the framework of landscape regions. This is accomplished by linking digital maps of landscape regions in GIS format to national databases (see Figure 22.3). It then becomes possible to examine the distribution of land uses, production, housing, etc., within landscape units at whatever scale is of interest. This development provides a new and powerful research and planning tool as well as the opportunity to validate the methodology and reality of landscape regions and their boundaries.

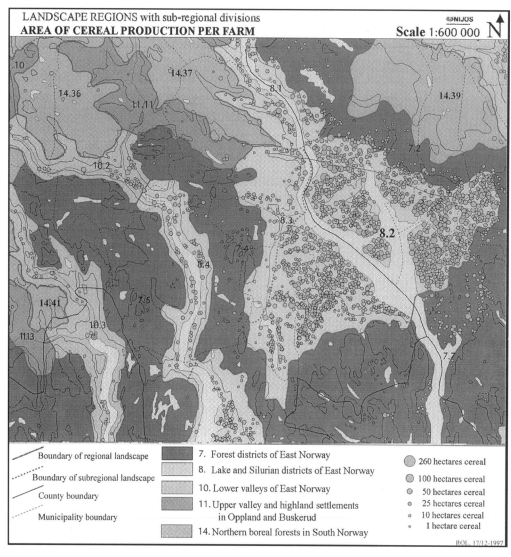

Figure 22.3 Map showing the cereal production per farm by linking grid references in the Norwegian farm databases with data from applications for agricultural subsidies in 1996. Production per farm is displayed to show the differences within and between landscape regions. Compilations of data for all types of production are used to assess the importance of farming systems for landscape character.

Potential markets for the landscape mapping system are in the management of national parks (Lykkja, 1998), assessing cultural heritage (Puschmann, 1996), local biodiversity plans, typology of cultural landscapes, and agricultural policy development (Nersten *et al.*, 1999). Pilot studies have used the methodology to identify landscape regions vulnerable to specific environmental stresses. The method can be used as input to environmental impact assessments or to plan the distribution of agri-environmental payments (Figure 22.1).

The main challenges ahead for the Norwegian system are to validate the landscape classes used at different scales and the utility of the landscape assessments in landscape planning and management.

22.7 General Discussion

Mapping and assessment projects aim to provide a better overview of landscape resources. Methods should be reviewed in the context of their purpose. Why we are mapping will determine the suitability of alternative approaches. It is unlikely that we will ever find a perfect method effective for all uses. Methods based on the numerical analysis of climate data have successfully identified zones suitable for different plant species in Australia (Austin and Myers, 1995), whereas approaches based on landscape perception have been valuable in identifying the landscape context and spatial relationships between iron-age burial mounds (Tilley, 1994; Gansum *et al.*, 1997). Several recent landscape character assessments attempt to combine the best of these contrasting approaches.

A list of points for consideration when assessing landscape classification and assessment based on the work of Brabyn (1996) and Stiles (1996) is presented in Table 22.2. Their arguments are combined and expanded to provide criteria for the validation of landscape assessment schemes and especially the classifications used to define mapping units. We recommend that methods be judged through consensus among landscape researchers and professionals in the context of their aims and against such criteria. In the European Environment Agency's Landscape Monograph (Wascher, 1999, p. 8) the difficulty of achieving this goal is summarised by the quote

> " *Both the classification and assessment methodology have been challenging tasks which require a high degree of theoretical groundwork as well as a clear practical orientation*".

Although it is increasingly common for landscape mapping schemes to be digitised into a GIS, few of these have, as yet, utilised the full analytical potential of GIS (Berry, 1993). In most cases GIS has been no more than the icing on the mapping cake. The next stage in the development of landscape mapping and assessment will see a greater use of the analytical powers of GIS in the classification of landscapes and in linking landscape regions to a wide range of other spatial data. Since all spatially referenced data can be used in conjunction with landscape mapping via GIS, new and exciting possibilities are likely to arise for linking qualitative and quantitative approaches.

Perhaps the greatest challenge to landscape mapping schemes in general is whether they can be validated. Even when we have clear objectives, several problem areas remain. For all forms of landscape assessment, we will have to cope with many anomalies including

Table 22.2 Suggested criteria for the validation of landscape mapping and assessment schemes adapted from Brabyn (1996) and Stiles (1996).

Criterion	Comments
Comprehensive spatial coverage	The classes must be exhaustive (can be expanded to include all landscape types), and mutually exclusive (classes must be clearly identifiable and not overlap).
Scale independent	Be capable of use at different scales. A hierarchical approach is best to cope with planning at different resolution and scales.
Widely applicable	Applicable from landscape to continental scales, but can cover both urban and rural landscapes, and can include present and historical landscapes.
Reproducible and consistent	Can be arrived at independently by different surveyors; has to produce repeatable results independent of surveyor.
Transparent	The principles can be described clearly and understood by users.
Recognisable	Map units should correspond closely to the perception of landscape form as a stable basis for attaching subjective values and evaluations.
Meaningful	The assessment of landscape should provide information on landscape function and structure.
Informative	Landscape assessment should provide information on relationships between landscape types (interactions, flows, responses to change processes) for prediction and asking the 'what if' question.
Reliable and authoritative	Major map classes based on a wide range of verifiable 'hard data'.
Widely accepted	The data used should be recognised and accepted by landscape researchers, planners and managers.
Flexible and open	Map classes must be flexible enough to incorporate new interests, values or priorities, rather than needing a new classification system.
Dynamic	Landscape assessment must recognise landscape dynamics (seasonal and longer) and be useful for prescriptive as well as descriptive purposes.

invisible landscape components. These may be the scene of a historical battle or the home of a famous artist or poet that give special significance and character to a place. Finally, we must accept that its usefulness will determine whether a particular mapping scheme survives or not. History has shown us that it is widely-used tools that survive, not necessarily the best ones. Continual interaction with user groups is, therefore, an essential step in method development if research efforts are to be implemented.

22.8 Conclusions

Many types of geographical information will be analysed with the help of GIS in the future. This is already having significant impacts on both research and planning. The role of landscape assessment in this development will rest on how useful individual schemes are in practice and whether users can understand and trust the methods.

A review of the Scottish method of landscape assessment (Environmental Resources

Limited, 1995) reflects on the uncertainties we face on standardisation, and what can and cannot be incorporated into GIS analysis. The future will see greater emphasis on developing quality control criteria against which methods can be judged. Such criteria will force an even sharper focus on the question of why we are mapping and our theoretical framework. Although the future will also offer new technological advances, we should avoid becoming fixated on the detailed advantages of one method over another, or what the latest software package can offer, as there will always be a newer and better version tomorrow. Instead, it will be more profitable to focus on the reasons for mapping landscapes and how success will be judged.

Finally, it must be remembered that landscapes are dynamic and always changing in response to natural and socio-economic driving forces. Mapped landscape units and their boundaries rapidly become fixed in the minds of planners and managers. The challenge is to map landscapes in a way that allows for dynamic changes in landscapes and in the reasons for mapping them.

Acknowledgements

We thank the editor Michael Usher for advice and two referees for valuable comments on an earlier draft. Thanks also go to Gro Jerpåsen for introducing us to landscape archaeology. The work is supported by the Norwegian Research Council.

References

Aldenderfer, M. and Maschner, H.D.G (Eds.) 1996. *Anthropology, Space and Geographic Information Systems.* Oxford, Oxford University Press.

Allen, K.M.S., Green, S.W. and Zubrow, E.B.W. (Eds.) 1990. *Interpreting Space: GIS and Archaeology.* London, Taylor and Francis.

Austin, M.P. and Myers, J.A. 1995. *Modelling of Landscape Patterns and Processes using Biological Data. Subproject 4: Real Data Case Study.* Canberra, CSIRO, Division of Wildlife and Ecology.

Berry, J.K. 1993. *Beyond Mapping: Concepts, Algorithms and Issues in GIS.* Fort Collins, GIS World Books.

Bruun, M. 1987. *Vurdering av landskapskvalitet.* Ås, Institute for Landscape Planning, Agricultural University of Norway.

Blankson, E.J. and Green, B.H. 1991. Use of landscape classification as an essential prerequisite to landscape evaluation. *Landscape and Urban Planning, 21*, 149-162.

Brabyn, L. 1996. *New Zealand Landscape Classification.* Christchurch, Canterprise.

CAG Consultants and Land use Consultants 1997. *Environmental Capital: A New Approach.* Unpublished report to Countryside Commission, English Heritage, English Nature and the Environment Agency. Cheltenham, Countryside Commission.

Carey, P.D., Preston, C.D., Hill, M.O. and Usher, M.B. 1995. An environmentally defined biogeographical zonation of Scotland designed to reflect species distributions. *Journal of Ecology, 83*, 833-845.

Countryside Commission 1993. *Landscape Assessment Guidance.* Cheltenham, Countryside Commission.

Dramstad, W.E., Olson, J. and Forman, R.T.T. 1996. *Landscape Ecology Principles in Landscape Architecture and Land-use Planning.* Harvard, Harvard University Press.

Elgersma, A. 1996. *Landscape Regions in Norway, with Sub-regional Divisions. Map scale 1:2000 000.* Ås, Norwegian Institute of Land Inventory.

Endter-Wada, J., Blahna, D., Krannich, R. and Brunson, M. 1998. A framework for understanding social science contributions to ecosystem management. *Ecological Applications, 8*, 891-904.

English Nature 1993. *Strategy for the 1990s: Natural Areas - Setting Nature Conservation Objectives.* Peterborough, English Nature.

Environmental Resources Limited 1995. *Overview of Landscape Assessment Methodology in Scotland.* Edinburgh, Scottish Natural Heritage.

Fagence, M. 1988. Geographically-referenced planning strategies to resolve potential conflict between environmental values and commercial interests in tourism development in environmentally sensitive areas. *Journal of Environmental Management,* **31**, 1-18.

Forman, R.T.T. 1995. *Land Mosaics: The Ecology of Landscapes and Regions.* Cambridge, Cambridge University Press.

Forman, R.T.T. and Godron, M. 1986. *Landscape Ecology.* New York, Wiley.

Fry, G. 1998. Changes in landscape structure and its impact on biodiversity and landscape values: a Norwegian perspective. In Dover, J. and Bunce, R. (Eds.) *Proceedings of the European International Association of Landscape Ecology Conference 'Key Concepts in Landscape Ecology',* Preston, International Association of Landscape Ecology. pp. 81-92.

Fry, G.L. and Sarlov-Herlin, I. 1995. Landscape design: how do we incorporate ecological, cultural and aesthetic values in landscape assessment and design principles? In Griffiths, G.H. (Ed.) *Landscape Ecology: Theory and Application,* Preston, International Association of Landscape Ecology. pp. 51-60.

Gansum, T., Jerpåsen G. and Keller, C. 1997. *Arkeologisk landskapsanalyse med visuelle metoder.* Stavanger, Arkeologisk Museum Stavanger.

Gauch, H.G. 1982. *Multivariate Analysis in Community Ecology.* Cambridge, Cambridge University Press.

Gjølberg, O. (Ed.) 1995. *Agricultural Policy and the Countryside. Proceedings of the Homenkollen Park Workshop.* Ås, Department of Economics and Social Sciences, Agricultural University of Norway.

Haines-Young, R., Green, D.R. and Cousins, S.H. (Eds.) 1993. *Landscape Ecology and GIS.* Taylor & Francis, London.

Herpsager, A.M. 1994. Landscape ecology and its potential application to planning. *Journal of Planning Literature,* **9**, 14-29.

Hertig, T. 1993. Nature experience in a transactional perspective. *Landscape and Urban Planning,* **25**, 17-36.

Hitier, P. 1998. The draft European Landscape Convention. *Naturopa,* **86**, 20-21.

Host, G.E., Polzer, P.L., Mladenhof, D.J., White, M.A. and Crow, T.R. 1996. A quantitative approach to developing regional ecosystem classifications. *Ecological Applications,* **6**, 608-618.

Kamfjord, G., Lykkja, H. and Puschmann, O. 1997. *Landskapet og reiselivsproduktet.* Ås, Norwegian Institute of Land Inventory.

Lawton, J. 1991. Ecology as she is done, and could be done. *Oikos,* **61**, 289-290.

Levine, J. and Landis, J.D. 1989. Geographic information systems for local planning. *Journal of the American Planning Association,* **55**, 209-220.

Llobera, M. 1996. Exploring the topography of mind: GIS, social space and archaeology. *Antiquity,* **70**, 612-622.

Lykkja, H. 1998. *Utarbeiding av forvaltningssoner for nasjonalparkar og deira nærområde.* Ås. Norwegian Institute of Land Inventory.

Macinnes, L. and Wickham-Jones, C.R. (Ed.). 1992. *All Natural Things; Archaeology and the Green Debate.* Oxford, Oxford Books.

Mitchell, B. 1989. *Geography and Resource Analysis (2nd Edition).* Harlow, Longman.

Nassauer, J.I. 1992. The appearance of landscape as matter of policy. *Landscape Ecology,* **6**, 239-250.

Nelson, J.G. and Serafin, R. 1992. Assessing biodiversity: a human ecological approach. *Ambio,* **21**, 212-218.

Nersten, N.K., Puschmann, O., Gudem, R. and Stokstad, G. 1999. *The Importance of Agriculture for the Norwegian Landscape.* Oslo, Norwegian Institute of Land Inventory and Norwegian Institute for Agricultural Economics Research.

Nordisk Ministerråd 1987. *Natural and Cultural Landscapes in Areal Planning. 1 Regional Classification of Landscapes (in Norwegian with English summary).* Moss, Nordisk Ministerråd.

Puschmann, O. 1996. *Et Landskaps Kulturkvaliteter.* Ås, Norwegian Institute of Land Inventory.

Puschmann, O. 1998. *Nasjonalt referansesystem for landskap.* Ås, Norwegian Institute of Land Inventory.

Ripple, W.J. (Ed.) 1987. *Geographic Information Systems for Resource Management: A Compendium.* New York, American Society for Photogrammetry and Remote Sensing.

Selman, P. 1993. Landscape ecology and countryside planning: vision, theory and practice. *Journal of Rural Studies,* **9**, 1-21.

Stanners, D. and Bourdeau, P. (Eds.) 1995. *Europe's Environment: The Dobris Report* . Copenhagen, European Environment Agency.

Stiles, R. 1996. The role of landscape assessment in monitoring land use change. In R. Jongman (Ed) *Ecological and Landscape Consequences of Land Use Change.* Tilburg, European Centre for Nature Conservation. pp. 196-211.

Tilley, C. 1994. *A Phenomenology of Landscape. Places, Paths and Monuments.* Oxford, Berg.

Wadsworth, R. and Treweek, J. 1999. *Geographical Information Systems for Ecology: An Introduction.* Harlow, Longman.

USDA Forest Service 1974. *The Visual Management System. Agricultural Handbook No. 462.* Washington, U.S. Government Printing Office.

Wagstaff, J.M. (Ed.) 1987. *Culture & Landscape: Geographical & Archaeological Perspectives.* London, Blackwell.

Wascher, D. (Ed.) 1999. *European Landscapes: Classification, Evaluation and Conservation.* Copenhagen, European Environment Agency.

Wickham, J.D. and Norton, D.J. 1994. Mapping and analyzing landscape patterns. *Landscape Ecology,* **9**, 7-23.

Williams, D.R. and Patterson, M.E. 1996. Environmental meaning and ecosystem management: perspectives from environmental psychology and human geography. *Society & Natural Resources,* **9**, 507-521.

Wright, R.G., Murray, M.P. and Merrill, T. 1998. Ecoregions as a level of ecological analysis. *Biological Conservation,* **86**, 207-213.

Zube, E.H. 1984. Themes in landscape assessment theory. *Landscape Journal,* **3**, 104-110.

INDEX

Most references are to **landscape** and **Scotland**, unless otherwise indicated. *Italics* indicate maps and illustrations. Plates are indicated by *pl.* before page numbers, for example: *pl. 25.* **Emboldened** page numbers indicate chapters.

Aberdeen: landscape character
  assessment 7, *8*, 172
  development planning *pl. 4*; 5,
    13, 18-21, 22
Aberdeenshire 7, *8*
  planning new housing 94-5, 96-7
accessibility 48-9
  and mapping remote areas 115-17,
    119, 120
aerial photography **101-11**, 164, 185
  flight path creation 106
  geographic image data 137
  identification of change 108-10
  relict landscapes *pl. 24-5*
  sequence editing 107-8
  see also geographic image data;
    satellite imagery
afforestation *see* forestry
Agenda 21: 43
AGLVs (Areas of Great Landscape
  Value) 15, 27, 62
agriculture
  and aerial photography 109
  Agricultural Census 180
  Agriculture Environment and
    Fisheries Department 3
  and cultural issues 177
  and development planning and LCA
    13, 14
  field pattern *pl. 25*; 165, 166
  and Historic Land Use Assessment
    166, 167-8
  and information technology 145
  and LANDMAP approach 80, 81
  and landscape character assessment
    3, 5
  and local distinctiveness in Sweden
    42-3, 44, 46, 51-2
  and Natural Heritage Zones 30
  Norway 190, *198*, 199
  and planning new housing 88, 92
  and satellite imagery 129
Aho, A.V. 117
Aitken, R. 34, 35, 38, 117, 118
Aldenderfer, M. 191, 193
Aldie *pl. 24*; 166
Allen, K.M.S. 191
Allt Mór valley 121
Alnarp Landscape Laboratory
  (Sweden) *pl. 6*; 46-7, 52
Alness 72

altitude 113, 117
'amenity' concept 2
Ammons, R.B. 35
Anderson, R. 75
Angus 90, 91
animals *see* flora and fauna
anthropogenic features *see* man-made
Appleton, J. 42
applying concepts of landscape
  character *see* associations; forest
  design guidance; LANDMAP
  approach; planning new housing;
  tourism, sustainable
AR (augmented reality) 156
ARC/INFO 39, 117
archaeology 19, 59, 164-5, 167-8
architecture 73, 92
Ardnamurchan 5
Ardverikie 7, *8*
Area Tourism Strategies 75
Area Tourist Boards 68, 71, 75
Areas of Great Landscape Value 15,
  27, 62
Areas of Regional Landscape
  Significance 94
Areas of Regional Scenic Significance
  15
Argyll 7, *8, 28-9*, 71
ARL (average radial length) 131
Arrochar Alps 37, *38*, 39
artificial intelligence 151, 153
ASH Consulting Group 7, 9
Aspinall, R.J. 101
associations triggered by specific
  landscape characteristics *pl. 17-18*;
  83-7
ATBs (Area Tourist Boards) 68
augmented reality 156
Austin, M.P. 191, 199
Australia *pl. 22*; 35, 121, 155, 199
Autonomous Agent modelling 157
average radial length 131
Aviemore 70-1, *116, 118-20, 122-4*
Avon, Loch *116, 118-20, 122-4*
Aylward, G. 153, 155
Ayrshire 7, *8*, 75

Badenoch 74
Balharry, D. 10, 26
Baltsavias, E.P. 103

Banff 7, *8*
Barnett, R. 68
barrier features in remote areas 113,
  118, *120*
base aspects in LANDMAP approach
  78
BasicScale digital maps 164
Baxter, C. 68
beauty *pl. 1*; 2-3
behavioural modelling 157, 158
Belgium *see* satellite imagery
Bell, S. xiv
  on forest design 9, 15, **57-65**
Bellamy, D. 121
Ben Alder 7, *8*
Ben Nevis 37, *38*, 39, 73
Benefield, C.E. 61
Bennett, A. xiv
  on monitoring and measuring
    landscape change 9, 172, **180-2**
Bennett, S.P. xiv
  on development planning **13-22**
Bergen, S.D. 153, 156
Berger, P. 102, 139, 153, 155
Bergson, H. 43-4
Berry, J.K. 199
biodiversity 30, 43
  Biodiversity Action Plan, UK 31-2
Birnie, R.V. xiv
  on monitoring and measuring
    landscape change **180-2**
Bishop, I.D. xiv
  on modelling the view **150-61**
Black Hill (USA) 49
Blairadam 166-7
Blankson, E.J. 191
Borders
  Historic Land Use Assessment *pl.
    25*; *165*, 166, 167
  landscape character assessment 7, *8*
  Natural Heritage Zone *28-9*
  planning new housing 91, 94
  tourism 75
Boster, R.S. 153
Boswell, J. 68
bottom upwards approach 48, 192
  see also local issues
boundaries 131, 191, 195
Bourdeau, P. 191, 192
Brabyn, L. 147, 191, 192, 199, 200
Bradbury, R. 121

Bramme, A. ix
Breadalbane *28-9*
Brinckerhoff Jackson, J. 42
Bristol (England) 184
Brooke, D. 128
'brownfield' sites 20
Brush, R.O. 153
Brussels (Belgium) 131-2
Bruun, M. 194
Bryden, D. xiv
   on sustainable tourism **66-77**
Buchan, N. v, xiv
   on landscape character assessment of
      Scotland **1-12**, 27, 43, 70, 192
Buchan 7, *8*
Bucht, E. 44
buffering approach to mapping
   remote areas 115, *116*
buildings *see* housing
Bullen, J.M. xiv
   on LANDMAP **78-82**
Bunce, R.G.H. 61
Burrough, P.A. 115
Bushyoff, G.J. 153

CAG Consultants 191
Cairngorms
   and aerial photography *see*
      Glenfeshie
   and cultural issues 177
   Historic Land Use Assessment *165*,
      169
   landscape character assessment 7, *8*,
      10
   Look Back Studies 180
   mapping remote areas 115, *116*,
      117, *118-20*, 121, *122-4*, 125
   National Park proposed 169
   Natural Heritage Zone *28-9*
   wild land 35, 37, *38*, 39
Caithness 7, *8*
Caledonian Forest 177
Campbell, L. xiv, 7
   on landscape in development
      planning **13-22**, 172
Canada 34, 35, 49
Cardiff (Wales) 79
Carey, P.D. 191
Carlson, A. 150
Carver, S. xiv
   on remote area mapping **112-26**
case studies 46
   see also Aberdeen; Dumfries and
      Galloway; Norway; Skrylle
CAVE (Cave Automatic Virtual
   Environment) 155
CC and CCS *see* Countryside
   Commission, etc.

CCW *see* Countryside Council for
   Wales
CD-ROMs 106, 140, 142
Central Region 91, 180
   Historic Land Use Assessment *165*,
      166-7
   landscape character assessment 7, *8*,
      9
changes
   future 5
   land use 6, 9
   and local distinctiveness in Sweden
      42, 44
   see also management and change
Character of England Map 147
character of landscape *see* landscape
   character assessment
Cherry, G. 88
Chrisman, N.R. 136
Cities Revealed 138, 142
Clackmannanshire 7, *8*, 9, 171, 172
classification systems, major 191-2
   see also ITE; LANDMAP; landscape
      character assessment
Cleish *165*, 166
climate and weather 3, 26, 58, 191
   and local distinctiveness in Sweden
      46, 51
   and remote areas 113, 118
Clwyd (Wales) 60
Clyde Valley and Firth 7, *8*
coal mining, open-cast 9, 167, 172
coastal areas 19, 30, 81, 185, 195
   see also islands
Cobham Resource Consultants 7
Collingwood, R.G. 45
Combined Aspect Areas 78, 81
Common Agricultural Policy 5, 174
Community Forests 61
Community Plans 174-5
computer 44, 185
   as support tool 151-8
   Virtual Reality 153, 155-7
   see also aerial photography;
      geographic image; information
      technology; view modelling
concepts of landscape character *see*
   applying concepts
conservation 41
   and LANDMAP approach 80-1
   and landscape character assessment
      3, 15-16
   and Natural Heritage Zones 25, 26
construction *see* housing and
   construction
Continuing Professional Development
   20
Cooper, A. 60

CORINE land cover database 130,
   180
Cornwall (England) 163
Cornwallis, G. 68
Cosgrove, D. 179
cost/push factors and remote areas
   113, 115, 118
costs 5, 139, 141-2, 183-4, 185-6
Council for Protection of Rural
   England 58
Countryside Agency 6
Countryside Character Programme
   (England) 61, 62, 64, 128
Countryside Commission and
   Countryside Commission for
   Scotland 3, 4, 95, 163, 180, 191
   and forest design guidance 57, 60,
      61-2
   and Historic Land Use Assessment
      163
   and landscape character assessment
      1, 4, 18-19, 21
   merger with Nature Conservancy
      Council for Scotland *see* SNH
   and planning new housing 96
   and planning and policy issues 174
Countryside Council for Wales 4, 57,
   145, 147
   see also LANDMAP approach
Countryside Design Summaries 96
Countryside Landscape Assessment 94
Countryside Survey 180, 181
Craik, K.H. 153
Crawford, D. 153
Creag Meagaidh 7, *8*
Cromarty 7, *8*
Cronon, W. 52
Crowe, N. 42
Crowe, Dame S. 58-9
Cuddy, S.M. 113
Cuillins of Skye 37, *38*, 39
cultural issues 4, 78, **176-9**, 190-1,
   200
   geneaology and history 176
   landscape change and mapping 177
   landscape character assessment
      programme 176, 177-9
   language and values 176-7
Cwm Berwyn forest (Wales) *102*,
   103-6, *109*, 110

damage to environment
   and LANDMAP approach 80-1
   and tourism 69, 73, 74-5
dams 75
Daniel, T.C. 153
Daniels, S. 179

databases and datasets 10, 27, 198
  access 182
  availability 183-4
  management 184-5
  see also computer; geographic image;
  GIS; landscape character assessment;
  technological developments
Deeside 73
DEMs (digital elevation models) 131
  and aerial photography 101, 103,
  104-5, 106, 109, 110
  and mapping remote areas 117,
  118-20, 122-4
Denmark 70
Department of the Environment,
  Transport and the Regions 62, 96,
  180, 184
Department of Health for Scotland
  88-9
Design of Forest Landscapes 59
Development Control Officers 90, 95
*Development in Countryside and Green
  Belts* 90
development planning 172, 173
  and housing 89, 96
  landscape character assessment used
  in 5, 10, **13-22**
  see also housing; Local Plans;
  Structure Plans
digital elevation models *see* DEMs
Dijkstra's algorithm 117-18
distance, linear *see* proximity
distinctiveness *see* local distinctiveness
Dixon, P. xiv
  on Historic Land Use assessment 10,
  162-9, 176
Dortmans, C. 128
Douglas, D.H. 113
drainage *see* hydrology
Dramstad, W. xiv
  on Norway **189-203**
drystone walls 72, 75
DTM xi, 141
Dufourmont, H. xiv
  on satellite imagery **127-53**
Dumfries Coastlands 14
Dumfries and Galloway: landscape
  character assessment 7, 8, 9, 172
  and development planning *pl. 3*; 5,
  13-18, 21-2
  and forest design guidance *pl. 11-12*;
  57, 62-4
Dumfries and Galloway Enterprise 14
Dumfries and Galloway Natural
  Heritage Zone *28-9*
Duncan, J. 178
Dunfermline 7, *8,* 172
Dunn, M.C. 153
dynamic perspective on local
  distinctiveness in Sweden 43-5

Dyson Bruce, L. xiv
  on Historic Land Use assessment
  162-9

Eastern Lowlands Natural Heritage
  Zone *28-9*
ecology 3, 25, 59, 73, 78, 147, 190
  and local distinctiveness in Sweden
  44, 51
  see also flora and fauna; vegetation
economy and tourism 67, 76
Edinburgh 51
Edling, N. 83
education 44
EDX geographic image data 137
Eigg 5
Elgersma, A. 194
Ellsworth, J.C. 153, 156
Else, R. 121
Elsner, G.H. 150
Endter-Wada, J. 192
England
  Countryside Character Programme
  61, 62, 64, 128
  forest design guidance 58, 61-2, 64
  Historic Land Use Assessment 163
  information technology 147
  planning new housing 96
  planning and policy issues 174
  technological developments 184,
  185
English Nature 62, 147, 191
English Partnerships 184
enhancement 16
Enterprise Companies 25
Enterprise Network 25
Environment and Heritage Service
  (Northern Ireland) 4
Environmental Capital Approach 173,
  174
Environmental Resources Limited
  191, 192, 200-1
Environmental Resources
  Management 5, 7, 63
Environmentally Sensitive Areas 71-2,
  180
ERDAS xi, 103
Eriksberg (Sweden) 50
ERM (Environmental Resources
  Management) 5, 7, 63
ESAs (Environmentally Sensitive
  Areas) 71-2, 180
Estonia 51
Ettleton Churchyard *pl. 25*
Europe and European Union 25
  Common Agricultural Policy 5, 174
  Directives 31-2, 102

Environment Agency 199
  Landscape Convention 25
  and satellite imagery 130
  visitors from 69-70, 73
  see also Belgium; Germany; Norway;
  Sweden
evaluated aspects in LANDMAP
  approach 78
Ewert, A.W. 112
experiments 47
Fabos, J.G. 153
Fagence, M. 191
Fairbairn, M. xiv
  on technological developments
  183-6
Falkirk *165*
Falun (Sweden) 83, 84
Farrell, R.H. 69
Faust, N.L. 102, 135
FC *see* Forestry Commission
Feimer, N.R. 153
Fenton, J. 34, 115
Ferguson McIlveen 7, 35
fieldwork 5, 19
Fife 7, *8,* 94, 172
Findlay, I. 68
Fines, K.D. 59
Fisher, E. 136
Fisher, P. 155
fitness for purpose of geographic
  image data 136-7
*Fitting New Housing Development into
  the Landscape* 91
Fitzpatrick Associates 73
Flanders (Belgium) *pl. 19*; 128, 129
Fletcher, S. 7
flora and fauna 78
  biodiversity 30, 43
  and housing 92
  and local distinctiveness in Sweden
  44, 46, 50, 52
  and tourism 72
  and wild land 34, 39
  see also ecology; vegetation
Flow Country Natural Heritage Zone
  28-9
Flowerdale hydro scheme 72
Flyvbjerg, B. 47
focus groups 79-81
footpaths and tracks 39, 109
  and mapping remote areas 113, 118,
  120
  and tourism 72, 73
Forest Authority 3, 60, 180
forest design guidance integrated with
  landscape character 9, 14-15, **57-62**
  in Dumfries and Galloway *pl. 11-12*;
  57, 62-4
Forest Enterprise 3

Forest Landscape Design Guidelines 57, 59, 63-4

forest, natural 30, 92, 109, 138
  historical 167
  and LANDMAP approach 80-1
  local distinctiveness in Sweden 42-3, 47-8, 49, 51-2
  and satellite imagery and spatial analysis tools 129, 131, 132, 133

forestry and afforestation 5, 9, 146, 185, 190
  and aerial photography 109-10
  and development planning and LCA 14-15, 19, 21
  and Historic Land Use Assessment 166, 167
  and local distinctiveness in Sweden 47, 48
  and tourism sustainability 72, 75
  see also forest design guidance

Forestry Commission
  and forest design guidance 57, 58, 59, 60, 61, 62-3, 64
  and information technology 146
  and landscape character assessment 9, 14, 15, 21

Forestry Frameworks 64
Forestry Subject Local Plan 14-15
Forman, R.T.T. 190
Forrest, S. 157
Foster, C. 49
Fournier, A. 156
Fragstats 130
framing view 155
France 69-70, 178
Fransson, L. xiv, 44, 49
Fredriksson, C. 48
Fritz, S. xiv
  on remote area mapping **112-26**
Fry, G. vi, xiv, 44
  on Norway **189-203**
funding 5, 139, 141-2
funicular, proposed 120-1, *123*
future 46-50, 185
  changes 5
  and local distinctiveness in Sweden 41, 43, 45

Galloway 14
  wild land 37, *38*
  see also Dumfries and Galloway
Gansum, T. 199
Gauch, H.G. 192
GenaMap software 162, 164
Genasys Ltd 162
geneaology and history 176
generality 50
geographic image data for landscape

visualisation, high resolution **135-43**
  access 140-1
  accuracy 141
  availability 138-9
  costs 139, 141-2
  management 139-40
  problem *pl. 21*; 136-7

Geographical Information Systems *see* GIS
GeoInformation Group xi, 140
geology 3, 26, 59-60, 78, 128, 192
geomorphology and topography 3, 59, 78, 164
  and Norway 191, 192, 195
  and satellite imagery 128, 133
Germany 69-70, 73, 178
Gifford, J. 121
Gill, A. 35
Gillespies 7
Gimblett, H.R. 157, 158
GIS (Geographical Information Systems) 183
  and aerial photography 110
  and cultural issues 176, 177, 178
  and forest design guidance 60
  and high resolution geographic image data 140, 141
  and Historic Land Use Assessment 162, 163, 164-5
  and information technology 146, 148
  and local distinctiveness in Sweden 41, 47-8, 53
  national dataset of landscape character types *pl. 2*; 6
  and Norway 189, 193, 198, 199, 200-1
  and satellite imagery and spatial analysis tools 128, 133
  and view modelling 151, 153, 155, 157-8
  and wild land 34, 35-40
  see also mapping remote areas
Gjölberg, O. 190
Glamorgan, Vale of (Wales) 79-80
Glasgow 7, *8*
Glen Lui 177
Glen Shiel Hillwalking Survey 70, 73
Glencoe 37, *38*
Glenfeshie *102*, 103-6, *107*, *108*, 109
Global Positioning System 103, 145, 185
Godlovitch, S. 150
Godron, M. 190
Gold, J. and M. 75
Gordon 172
Gourlay, D. 72
GPS *see* Global Positioning System
gradient *see* slope

Graf, K.C. 153, 155
Grampians 91
Grangemouth 7, *8*
grazed landscape 49-50, 51, 52, 166
Great North Road 68
Green, B.H. 191
Green Belts 18, 20, 90
'greenfield' sites 18, 20, 81
ground surface 113, 118
guidance 174
  see also forest design guidance
guidebooks 68
guidelines
  Forest Landscape Design 57, 59, 63-4
  Guidelines for Landscape and Visual Impact Assessment 6
  Lowland Landscape Design Guidelines 61
  National Planning Policy Guidelines 10, 90, 172, 174
Gulinck, H. xiv
  on satellite imagery and spatial analysis tools **127-53**
Gustavsson, R. ix, xiv
  on local distinctiveness in Sweden **41-54**
Gwynedd (Wales) 79

Habitats Directive, EC 102
Habron, D. xiv
  on wild land **34-40**
Hadrian, D.R. 155
Hägerhäll, C.M. xiv
  on associations triggered by landscape types **83-7**
Haines-Young, R. 189
Hamill, L. 144-5
Hargrove, E. 150
Harvey, R. 181
Hawkins, D. 147
Hawkins, R. 67
head-mounted displays (in VR) 155, 156
Hebrides 7, *8*, *28-9*, 37, *38*
Hendee, C.J. 114, 121
Hendrix, W.G. 153
heritage *see* Historic Land Use; Natural Heritage Zones
Herpsager, A.M. 189, 190
Herries, J. 70, 73
Herrington, J. 155
Hertig, T. 191
Hester, A.J. 109
Hiebeler, D. 157
high resolution data *see* geographic image
Highland and Islands Enterprise 72

Highlands
  and cultural issues 177
  Historic Land Use Assessment *165*
  Natural Heritage Zones *28-9*
  planning new housing 90, 91
  and tourism sustainability 71, 75
  Visitor Survey 68, 72, 74
  wild land 34, 37, *38*, 39
  see also Cairngorms; mapping remote
    areas
Hingley, R. xv
  on Historic Land Use assessment
    162-9, 176
Historic Land Use Assessment 10,
  162-9, 176
  application 165-9
  methodology 163-5
  objectives 162-3
Historic Scotland 3, 10, 162, 163,
  169, 177
history 4, 78, 128, 176
  and aerial photography *108, 109,*
    110
  time view in Sweden 43-5, 49-50,
    53
Hitier, P. 190
HLA *see* Historic Land Use
  Assessment
HMDs *see* head-mounted displays
Hollenhorst, S.J. 112
Hoskins, W.G. 178
Host, G.E. 191
housing and construction 5, 81, 138,
  198
  architecture 73, 92
  and satellite imagery 128, 129, 131,
    132
  urban 18, 19
  in wild land 39
  see also under planning
Howard, J.A. 101
HS *see* Historic Scotland
HTML 140
Hughes, R. xv
  on landscape character assessment of
    Scotland v, **1-12**, 27, 43, 70, 192
Hulse, D.W. 153, 155
humans and landscape *see* individual;
  man-made
Hunter, J. 68, 73
Husserl, E. 44
hydrology, river systems and drainage
  129, 195
  and landscape character assessment
    7, *8*, 39, 71, 78
  and mapping remote areas 113, *118*

IACS (Integrated Administration and
  Control System) 180

iconography 178-9
identification criterion in geographic
  image data 137
IFS (Indicative Forestry Strategies) 59,
  60, 62, 171
Ihse, M. 44
IKONOS satellite 142
image data 136-7
image data, geographic *see* geographic
  image
images of landscape for tourists 69-70
Indians, native American 49
Indicative Forestry Strategies 59, 60,
  62, 171
individual perceptions 2, 3, *132*
  and Norway 190, 192
  of remote areas 113 *14*
  and tourism sustainability 66-7,
    69-70, 72
  and visualization *see* view modelling
  see also LANDMAP; subjectivity;
    wild land
industry 81, 90, 166, 185, 198
  coal-mining 9, 167, 172
  and tourism sustainability 72, 75
  see also forestry; mineral
information technology and landscape
  planning **144-9**
  in England and Scotland 146-7
  LANDMAP in Wales 146, 147
  requirements 147-8
  Visual Impact Analysis 245
informative aspects in LANDMAP
  approach 78-9
Ingelög, T. xiv, 44, 52
Institute of Environmental Assessment
  6
Institute of Terrestrial Ecology *see* ITE
Integrated Administration and
  Control System 180
integrated and interactive systems
  157-8
Inter-Agency Working Group 4
Internet *see* Web and Internet
intervisibility 114
inventory *see* landscape character
  assessment
Inventory of Garden and Designed
  Landscapes 27
Inverness 7, *8*
Ireland 73
IRS-IC satellite
  geographic image data 137
islands 5
  Hebrides 37, *38*
  Natural Heritage Zone *28-9*
  Rum 10, 37, *38*
  tourism 71, 72
  see also Orkney; Shetland; Skye

ITE (Institute of Terrestrial Ecology)
  57, 61, 180

Janssen, R. 124
Jepson, W. 156
Jirkhill *pl. 25*
Johnson, H.B. 155
Johnson, S. 68
Johnston, A. 7
Jones, A.C. xv
  on geographic image data **135-43**
Jones, T. 157

Kamfjord, G. 194, 195
Kaplan, R. and S. 83, 153
Kemp, K.K. 103
'key-stone elements' 46
Kierkegaard, S.A. 46
Kinross-shire 7, 8, 94, 172
Kirkpatrick, J.B. 155
Kliskey, A.D. 35
Knoydart 37, *38*, 39
Koch, N.E. 49
Kohlin, H.-J. 102
Krokshult (Smaland, Sweden) *pl.
  15-16*; 84, *85*, 86-7
Kuiper, J. 139

laboratory, landscape *pl. 6*; 46-7, 52
Lairig Ghru 121
Lake District (England) 58
land cover data sets 129-30
Land Cover Map of ITE 57, 61, 180
Land Cover of Scotland (Macaulay) 5,
  145, 164, 180
Land for Housing 90
land use 78, 192
  changes 6, 9
  defined 129
  and forest design guidance 59-60
  see also Historic Land Use
Land Use Consultants 5, 7, 14, 27,
  173, 174, 191
landfill and waste disposal 19
Landform Panorama 137
landforms *see* geomorphology
Landis, J.D. 189
LANDMAP approach (Wales) **78-82**
  and forest design guidance 60, 61,
    64
  and information technology 146,
    147
  photographs *pl. 13-14*; 79-81
  public perception 78-9
Landsat data 129-30, 137
Landscape Assessment Guidance 95

landscape change and role of tourism 74-6
landscape character assessment **1-12**
  applications 6-10
  concepts *see* applying concepts
  and cultural issues 176, 177-9
  in development planning **13-22**
  examples *pl. 1-4*
  and forest design guidance 62-3
  and Historic Land Use Assessment 162-3, 169
  and monitoring 181
  outputs 6
  and planning new housing 89, 94-6
  and planning and policy issues 171-4
  remoteness 113-14
  and satellite imagery 128, 130-2
  see also forest design guidance; local distinctiveness; Natural Heritage Zones; wild land
landscape damaged *see* damage; man-made
Landscape Design Guidance for Forests and Woodlands *pl. 11-12*; 14-15
Landscape Group, Advisory Services 7
Landscape Institute 6
Landscape Monograph 199
Landscape Strategy for Aberdeen 21
landscaping 20
Lange, E. 153, 155
Langmuir, E. 117
language 127
  place names 42, 45, 176, 177
  and values 176-7
Lanius, U.F. xi
Lau, D.C. 144
Law, A. xv
  on technological developments **183-6**
Lawton, J. 190
LCA *see* landscape character assessment
Leadbeater, S. xv
  on monitoring and measuring landscape change **180-2**
Lee, F. xv, 59, 61
  on cultural issues **176-9**
legislation 2, 24, 25, 31, 88, 114
Leopold, A. 114
Lesslie, R.G. 35, 68, 121
Levine, J. 189
Lewis 37, *38*
Liddesdale *pl. 25*; *165*, 166, 167
Liggett, R. 156
Lindgren, A. 86
Lindquist, B. 50
linear access in remote areas *119*, *120*

linear distance *see* proximity
Linklater, M. 1121
literature 4, 68
Llobera, M. 193
Llŷn (Wales) 79
Local Authorities 25, 27, 184
  and information technology 146, 147
  and landscape character assessment 5-6, 9
  and planning and policy issues 172, 173, 174, 175
local distinctiveness in Sweden **41-54**
  future 46-50
  reference landscapes 50-2
  time perspective 43-5
Local Forestry Frameworks 14-15
local issues and perspectives 25, 37, 174, 177, 195
Local Plans 172, 173, 195
  Aberdeen 19-20
  Aberdeenshire 94-5, 96-7
  Dumfries and Galloway 14-15
  Moray 92
  and new housing 91-2, 94
Loch Einich *116*, *118-20*, *122-4*
Loch Lomond
  landscape character assessment 7, *8*
  National Park proposed 169
  tourism 68, 71, 72
Loch Morlich 121
Lochaber 5, 7, *8*, *28-9*
Lochalsh 7, *8*, 11
Lochnagar hillwalkers 73
Lofoten (Norway) 195
Loh, D.K. 135
'long walk in' 114, 115
Lothians 7, *8*
Lumb, A. xv
  on planning and policy issues **171-5**
Lykkja, H. 199

Macaulay Land Use Research Institute v-vi, 145, 164, 180
McCullagh, M.J. 136
McDonnell, R.A. 103
McGarigal, K. 130
MacGregor, A.A. 68
Machars 14
MacInnes, L. xv, 190
  on cultural issues **176-9**
Mackey, E.C. vi, xv
  on monitoring and measuring landscape change **180-2**
MacLellan, R. 69
McNeish, C. 121
MacPherson, J. 68

Macpherson Research 70, 73
Mainland (Orkney) *165*
Malmö (Sweden) 47, 84, 85-7
man-made features 2, 72, 75, 163
  lack of *see* remoteness; wild land
  see also agriculture; associations; damage; forestry; housing; industry; mineral; recreation; tourism
management and change 42, 191
  monitoring 145, **180-2**
  see also cultural issues; Norway; planning and policy; technological developments
management strategies 10
MAP software 115
mapping
  remote areas using GIS **112-26**
    accessibility 115-20
    potential applications 120-5
    remoteness as index of landscape character 113-14
  see also classification systems; landscape character assessment; Norway; Ordnance Survey
Mar Lodge landscape character assessment 7, *8*, 10
marketing *see* promotion
Marks, B. 130
MARS (Monitoring Agriculture with Remote Sensing) 180
Maschner, H.D.G. 191, 193
Maslen, M. 121
Mather, A. 73
Maxwell, T.J. vi, xv
MCE (multi-criteria evaluation) 121, 124
measurement 137
  of landscape change 9, **180-2**
'memory, landscape' 44
Mencius 144
Mendel, L.C. 155
mental landscape 45
Merlewood *see* ITE
Merriam, I.C. 35
Microsoft 142, 164
Middleton, V.T.C. 67
Miller, D.R. xv, 103, 145
  on aerial photography **101-11**
mineral working 5, 9, 14, 72, 88, 167, 172
Minerals Subject Local Plan 14
Minnesota (USA) 49
Minto, P. xv
  on information technology and planning **144-9**
Mitchell, B. 191, 192
MLURI *see* Macaulay Land Use Research Institute
models *see* techniques and models

Moir, J. x, xv
  on housing planning **88 98**
Monadhliath 37, *38*
Monitoring Agriculture with Remote
  Sensing (MARS) 180
monitoring and measuring landscape
  change 145, **180-2**
Moray and Moray Firth
  landscape character assessment 7, *8*
  Natural Heritage Zones *29*
  planning new housing 92, *93*, 94
morphological measurements, radial
  130-3
Mountaineering Council of Scotland
  37
mountaineers and hill-walkers 70, 72
  access *see* mapping remote areas
  rules 115-16, 118
  and wild land 35, 37, *38*, 39
MrSID format 140
Muir, M. 178
multi-criteria evaluation 121, 124
Myers, J.A. 191, 199
Myklestad, E. 153
mythology 176

Nairn 7, *8*
Naismith, W.W.: Rule 115, 117-18
Nash, R. 114, 121, 124
Nassauer, J.I. 190
National Assessments 25
National Countryside Monitoring
  Scheme vi, 180
National Forest (England) 61
National Heritage Resource
  Assessment 173
National Interest Mapping Service
  Agreement 183, 184, 185
National Land Information System
  184
National Land Use Database for
  England and Wales 184
National Landscape Assessment and
  NHZs 26, 27, 30-2
National Mapping Agency *see*
  Ordnance Survey
National Monuments Record of
  Scotland 163-5
National Nature Reserve 10, 71
National Parks 49, 180, 199
  proposed 39, 125, 162, 169
National Prospectuses and NHZs 25,
  30-1
National Scenic Areas 62, 71, 180
  and landscape character assessment
  6, 10, 15, 16, *17*, 27
National Street Gazetteer 184
National Topographic Database 184

National Tourist Routes 71
National Trust for Scotland 10
National Wilderness Inventory,
  Australian 121
Natural Areas (England) 62
'natural beauty' *pl. 1*; 2, 3
Natural Heritage of Scotland 11
Natural Heritage (Scotland) Act
  (1991) 2, 24
Natural Heritage Zones programme
  10, **23-33**, 178
  maps *28-9*
  National Landscape Assessment 27, 30
  National Prospectuses 25, 30-1
  Zonal Prospectuses 25, 31-2
naturalness 35
  see also remoteness; wild land
Nature Conservancy Council for
  Scotland merged with CCS *see* SNH
NCMS (National Countryside
  Monitoring Scheme) 180
Nelson, J.G. 190
Nersten, N.K. 199
Netherlands 70, 127
network analysis 113, 115
New Houses in the Country 89
'New Map of England' 61-2
New Zealand 35, 121, 147
Newcastleton *pl. 25*; 167
NHRA (National Heritage Resource
  Assessment) 173
NHZs *see* Natural Heritage Zones
Nicol, I. xv, 5, 7
  on landscape in development
  planning **13-22**, 172
NIJOS (Norwegian Institute for Land
  Inventory) 193-4
NIMSA (National Interest Mapping
  Service Agreement) 183, 184, 185
NLIS (National Land Information
  System) 184
NNR (National Nature Reserve) 10
North America 34, 35, 49-50
  see also United States
North East Coastal Plain Natural
  Heritage Zone *28-9*
North East Glens Natural Heritage
  Zone *28-9*
North Sea oil and gas 18
North West Seaboard Natural
  Heritage Zone *28-9*
Northern Highlands Natural Heritage
  Zone *28-9*
Northern Ireland 4
Norton, D.J. 189
Norway: geographic information for
  research and policy **189-203**
  landscape classifications 190-3
  mapping system 193-200

Norwegian Institute for Land
  Inventory 193-4
Nova Scotia (Canada) 49
Novar Wind Farm 72
NPPGs *see* National Planning Policy
NSAs *see* National Scenic Areas
NTB (National Topographic
  Database) 184
NTRs (National Tourist Routes) 71
numerical solution 144-5
NWI *see* National Wilderness
  Inventory

object
  -based concepts 41
  data in geographic image data 136
objectivity 2-3, 26, 177, 193
Oelschlaeger, M. 114
off-road travel *see* mapping remote
  areas
Oh, K. 153
Ohio (USA) 49
Olwig, K.R. 52
Ordnance Survey 5
  and aerial photography 101, 103-4,
  110
  and cultural issues 177
  geographic image data 137
  and Historic Land Use Assessment
  163-4
  Land Line 103, 180, 183
  and monitoring 180
  and technological developments
  183-6
Original Object Name Books 177
Orkney Islands
  Historic Land Use Assessment *165*
  landscape character assessment 7, *8*,
  9
  Natural Heritage Zone *28-9*
Orland, B. 102, 153, 156
OrthoMax 103
Osborne, J.R. 35
Our Common Future 24
Outdoor Recreation Resources Review
  Committee 114
Oxhagen (Scania, Sweden) *pl. 7-9*; 52

Pan-European Biological and
  Landscape Strategy 25
PANs (Planning Advice Notes) 10
pastoral landscape 49-50, *51*, 52, 166
Path Finder Areas 174
Patterson, M.E. 191
pedestrians *see* mountaineers
Pennines (England) 174
Pennsylvania (USA) 49

Pentland and Orkney Natural
  Heritage Zone *28-9*
perception *see* individual perceptions
Perth 94, 172
Peterken, G. 50
philosophy, 'traditional' 43-4
photography 185
  LANDMAP approach *pl. 13-14*;
    79-81
  and view modelling 151, 153
  wild land *pl. 5*; 35-9
  see also aerial photography; satellite
physical features of landscape 3
  see also climate; geology;
    geomorphology; soils; vegetation
place
  concept 41-2, 48
names 42, 45, 176, 177
sense of *see* sustainable *under* tourism
planning
  applications 19, 93, 95
  landscape character assessment used
    in **13-22**
  and local distinctiveness in Sweden
    43, 44
  new housing in countryside **88-98**
  examples *pl. 17-18*, *93*, 96
  landscape character assessments
    used 94-6
  policy context 89-94
  Planning Advice Notes 10, 89, 90-4,
    174
  and policy issues **171-5**
  and landscape character assessment
    171-4
  Stirling and Clackmannanshire 171,
    172
  and tourism, sustainable 69-74
  see also information technology
plantations *see* forestry
policy
  European 5, 174
  guidelines 10, 90, 172, 174
  legislation 2, 24, 25, 31, 88, 114
  and planning *see under* planning
Popper, K.R. 45
Portable Revision Interactive Survey
  Module 185
Positive Aesthetics 150
power supply lines and
  telecommunications masts 14, 71,
  72
presentation criterion in geographic
  image data 137
Price, G. 60
'primary wilderness' 34-5
PRISM (Portable Revision Interactive
  Survey Module) 185
promotion 11, 69, 73

proximity in mapping remote areas
  113, 115, *116*, *122*
Public Inquiry 19
public perception *see* individual
publicity 11, 69, 73
Puschmann, O. xv
  on Norway **189-203**

quality 44
QuickTime 106

Rackham, O. 50
radial morphological measurements
  pl. 19; 130-3
railways 113, 138
Rannoch Moor 37, *38*
RCAHMS *see* Royal Commission
recreation *pl. 1*; 19, 58, 88, 112
  and Natural Heritage Zones 25, 30
  and view modelling 157-8
  see also mapping remote areas;
    mountaineers; tourism
red-painted cottages *see* associations
reference landscapes in Sweden *pl.
  7-9*; 41, 50-2, 53
  historical 41
  key 46
Regional Character Areas 14
Regional Scenic Areas and
  Designations 14, 16, *17*
regionalism 194-5
*Register of Landscapes of Outstanding
  Historic Interest in Wales* 163
remote sensing 129-30, 151, 153, 180
remoteness 35
  and accessibility 115-20
  as index of landscape character
    113-14
  see also mapping remote areas; wild
    land
Renewable Energy Subject Local Plan
  14
Rhins 14
Rice, D. x, xv
  on housing planning **88-98**
Richards, J. 7
Rietveld, P. 124
Ripple, W.J. 189
river systems *see* hydrology
Rizell, M. 46
RMS (root mean squares) 104-5
roads 14, 81, 92, 128, 138, 185
  and mapping remote areas 113, *118*
  in Sweden 52
  and tourism 71
  see also footpaths

Rohrmann, B. 156
Rolston, H. 155
romanticism 68
root mean squares 104-5
Ross and Cromarty 7, *8*
Rowe, J. 135
Royal Commission on Ancient and
  Historical Monuments of Scotland
  162, 163, 167, 169
Rum 10, 37, *38*
Rural Development Strategies 174
Rural Guidance 174
Russell, J.A. xi

St Andrews 94, *165*, 172
Sanday *165*
Saremba, J. 35
Sarlöv-Herlin, I. 44
satellite imagery and spatial analysis
  tools 135, 138
  in Belgium **127-53**
  land cover data sets 129-30
  linkage to real world landscape
    character appraisal 131-2
  morphological landscape types 131
  spatial indices of landscape character
    130-1
  Global Positioning System 103, 145,
    185
  see also aerial photography;
    geographic image data
Scania (Sweden) *see* Oxhagen
Schifferl, E. 84
Scott, J.M. 129
Scott, Sir Walter 68
Scottish Development Department
  15, 88-9, 90
Scottish Enterprise vi, 67, 69, 71, 74
Scottish Environment Protection
  Agency 25
Scottish Mountaineering Club 117
Scottish Natural Heritage *see* SNH
Scottish Office 3, 10, 39, 180
  and planning new housing 88-9,
    90-4, 95, 96
  and planning and policy issues
    174-5
Scottish Tourism Co-ordination
  Group 69
Scottish Tourism Research Unit 70
Scottish Tourist Board 66, 68
Scottish and Wildlife Countryside
  Link 37
'secondary wilderness' 34-5
Sedgewick, R. 113
Selander, S. 50
Selman, P. 75, 190
sensitivity of landscape 16

SEPA (Scottish Environment
Protection Agency) 25
Serafin, R. 190
service level agreements 184
settlements 3, 78, 172
  see also housing; urban areas
Sewel, J. xiii
Shafer, E.L. 153
Shang, H.-D. 156
Shepherd, N. xv
  on forest design 9, 15, **57-65**
Shetland Isles
  Amenity Trust 3
  landscape character assessment 7, *8*
  Natural Heritage Zone *28 9*
Shieldaig inquiry 6
Shuttleworth, S. 156
sites of scientific importance 92
*Siting and Design of New Housing in
  Countryside* 89, 90-4
Skånes, H. 44
Skrylle Recreational Forest project
  (Sweden) 47-9, 52
Skye
  Historic Land Use Assessment *165,
    167, 168*
  landscape character assessment 7, *8,*
    11
  tourism 73
  wild land of Cuillins 37, *38*, 39
SLAs (service level agreements) 184
Slee, W. 72
slope/gradient 39, 113, 115, 117
Småland (Sweden) *see* Krokshult
Smallman, T. 68
Smardon, R.C. 150
Smout, T.C. xv
  on cultural issues **176-9**
SNH (Scottish Natural Heritage)
  1-11, 18-19, 43
  established (1992) 3
  and forest design guidance 57, 63
  and information technology 146-7
  and planning new housing 94, 96-7
  see also landscape character
    assessment; Natural Heritage Zones
Snogeholm (Sweden) 47
soils 3
  and forest design guidance 58, 59
  and satellite imagery 128, 129, 133
solitude 73
Solway Coast Natural Heritage Zone
  28-9
Söndergaard Jensen, F. 49
Sorte, G. 42
Southampton 185
Southern Uplands 13, 14
Space Imaging 142
spatial analysis tools *see* satellite
  imagery

spatial indices of landscape character
  130-1
Spatial Layout for Benelux 128
Spey, River 71
Spin-2 satellite 142
sports 70
SPOT satellite data 129-30, 137
spruce plantations *see* forestry
Staffordshire 60
Stanners, D. 191, 192
Stanton, C. 7
Stas, I. xv
  on satellite imagery and spatial
    analysis tools **127-53**
static perspective on local
  distinctiveness in Sweden 43-5
STB *see* Scottish Tourist Board
Steinitz, C. 155
Stevenson, J. xv
  on Historic Land Use assessment
    162-9
Stiles, R. 191, 199, 200
Stirling 71
  landscape character assessment 7, *8,*
    9
  planning and policy issues 171, 172
story telling 49
Stoyanoff, N.J. 84
Strathclyde 62
Strathspey 74
Structure Plans 9, 173
  Dumfries and Galloway 14-16, *17*
  Flanders 128
  Stirling and Clackmannanshire 171,
    172
subjectivity 177, 181, 192-3
  see also individual perceptions
sustainable development/sustainability
  9, 24, 182
  and local distinctiveness in Sweden
    43, 45
  see also under tourism
Sutherland 7, *8*, 74
Sweden *see* associations; local
  distinctiveness

Tang, H. 158
Tasmania 155
Taylor, S.G. 121
Tayside 7, *8*
techniques and models *see* aerial
  photography; geographic image;
  Historic Land Use; information
  technology; mapping remote areas;
  satellite imagery; view modelling
technological developments **183-6**,
  189-90
  data management 184-5

dataset availability 183-4
future 185
see also computer; geographic image;
  GIS; satellite
Tektronix terminals 151
terrain data in geographic image data
  136-7
Terraserver 142
Thayer, R.L. 153
Thin, F. xv
  on Natural Heritage Zones 10,
    23-33
Thompson, D.B.A. 68
three-dimensional computer models
  pl. 21; 155, 157, 185
  see also aerial photography;
    geographic image data
TIFF imagery 139
Tilley, C. 192, 199
timber *see* forestry
time
  taken to travel in remote areas 113,
    115, 116, 118
  see also history
Tjärö (Sweden) *pl. 10*; 52
Toftanäs (Malmö, Sweden) 47
Tomlin, D. 115
tools *see* techniques and models
top downwards approach 48, 192
  see also legislation; planning; policy
topography *see* geomorphology
tourism 13, 14, 88, 158, 182, 195
  sustainable **66-77**
    existing relationships 68-9
    landscape change and role of 74-6
    landscape planning 69-74
  see also recreation
*Towards a Development Strategy for
  Rural Scotland* 174
Town and Country Act (1947) 88
transport 14
  and mapping remote areas 113, *118*
  public 9
  in Sweden 52
  and tourism 71, 75
  see also roads
Tranter's Correction 117
Treweek, J. 189, 193
Trossachs
  landscape character assessment 7, *8*
  National Park proposed 169
  tourism 71
  wild land 37, *38*, 39
Trottenish *165, 167, 168*
Turnbull Jeffrey Partnership 7, 10
Turnbull, M. 153, 155
Turner, B.L. 129
Turner, J.M.W. 58
Turner, T. 44

TWINSPAN analysis 60, 61-2
Tyldesley, D. xv, 7, 94
  on planning and policy issues **171-5**
Tyndrum 73
types of landscape *see* landscape
  character assessment

UK Forestry Standard 59
United Nations 24
United States 142
  and cultural issues 178
  geographic image data 137-8
  scenic planning 194
  and Sweden compared 49-50
  view modelling 155
  wild land and remote areas 34, 35,
    114
Unwin, K.I. 153
Uplands Glens landscape character
  type *pl. 11-12*
urban areas and urbanisation 7, *8*, 14,
  30, 138, 166
  and housing planning 88-9
  and satellite imagery 128, 129, 132,
    133
  in Sweden 43
  see also Aberdeen; Stirling
Urry, J. 69
Usher, M.B. vi, xv, 10, 26

validation 199-200
values 43, 176-7
vegetation 3, 78, 90, 146, 155
  in Australia 199
  and mapping remote areas 113, *120*
  and Norway 191, 192
  and satellite imagery 129, 131, 132,
    133
  and wild land 34, 39
  see also flora and fauna; forest,
    natural; forestry
VEGETATION data 130
VIA (Visual Impact Analysis) 147,
  245
video fly-throughs *pl. 22*; 101, 106,
  107, 108-10
view
  modelling **150-61**
    computer as support tool 151-8
    research 151, 152-4
  and tourism 70, 72
  see also vision/visual
viewshed 39, 130, 155
Village Design 96
Vining, J. 156
Virtual Reality 153, 155-7

vision/visual
  impacts 16, 19
    Visual Impact Analysis 147, 245
    Visual Impact Assessment 6
  influence, sphere of 90
  visibility 39, 114
  visualization 179
    see also aerial photography;
  geographic image data; view
VistaPro 155
Vlaamse Regering 128
VR (Virtual Reality) 153, 155-7
VRML (Virtual Reality Modelling
  Language) 157

Wadsworth, R. 189, 193
Wager, J.A. 153
Wales 4, 145
  forest design guidance 57, 60, 61,
    62, 64
  Historic Land Use Assessment 163
  see also Cwm Berwyn; LANDMAP
walkers *see* mountaineers
Wallonia (Belgium) 127-8
walls 72, 75
Ward Thompson, C. 42
Warnock, S. 60
Warwickshire Landscape Assessment
  study 59-60
Wascher, D. 190, 191, 192, 199
waste disposal *see* landfill
water *see* hydrology
Waternish *165*, 167, *168*
Watt, A. x, xv
  on housing planning **88-98**
Watzek, K.A. 153, 156
weather *see* climate
Web and Internet *pl. 23*; 141, 142,
  147, 156-7, 184, 185
Welsh Lanscape Partnership Group 78
West Central Belt Natural Heritage
  Zone *28-9*
Wester Ross 35, 37, *38*, 39, 72
Western Highlands Natural Heritage
  Zone *28-9*
Western Isles 7, *8*, *28-9*, 37, *38*
Western Seaboard Natural Heritage
  Zone *28-9*
Wherrett, J.R. 156
Whitney, G. 50
Wickham, J.D. 189
Wickham-Jones, C.R. 190
Wigtown Machairs and Solway Coast
  Natural Heritage Zone *28-9*
wild land, landscape attributes of
  34-40, 73, 115
  defined 114
  photographs *pl.* 5; 35-9

see also remoteness
Wilderness Act (US, 1964) 114
wildlife *see* flora and fauna
Williams, D.R. 191
Williams, R. 178
wind energy and windfarms 9, 10, 71,
  145
  and development planning and LCA
    14, 15-18, 21
  Wind Energy Strategy 14, 15-18, *17*
  Windfarm Strategy 15
wind and walkers 113
Windy Standards windfarm 16
Withers, C. xv
  on cultural issues **176-9**
Wittrock, S. 45
Woelfel, J. 84
wooden fences *see* associations
Woodland Grant Scheme 15, 59, 63
Woodland Inventory 180
Wordsworth, D. and W. 68
World Commission on Environment
  and Development 24
World Wide Web *see* Web
Wright, R.G. 191

Xiang, W.-N. 155

Yates, P.M. 158
Yellowstone (USA) 49

Zewe, R. 102
zonal concept
  in mapping remote areas *124*, 125
  Zonal Prospectuses 25, 31-2
  Zone of Visual Influence 145
  see also Natural Heritage Zones
Zube, E.H. 192

Printed for The Stationery Office, 11/99
c8, J97621, cnn 001285